Stat~~istics in~~ ~~Ecological~~ ~~biology~~

Ecological and Environmental Toxicology Series

Series Editors

Jason M. Weeks
Institute of
Terrestrial Ecology,
Monks Wood, UK

Sheila O'Hare
Scientific Editor
Hertfordshire, UK

Barnett Rattner
Patuxent Wildlife
Research Center,
Laurel, MD, USA

The fields of environmental toxicology, ecological toxicology and ecotoxicology are rapidly expanding areas of research within the international scientific community. This explosion of interest demands comprehensive and up-to-date information that is easily accessible both to professionals and to an increasing number of students with an interest in these subject areas.

Books in the series will cover a diverse range of relevant topics ranging from taxonomically based handbooks of ecotoxicology to current aspects of international regulatory affairs. Publications will serve the needs of undergraduate and postgraduate students, academics and professionals with an interest in these developing subject areas.

The Series Editors will be pleased to consider suggestions and proposals from prospective authors or editors in respect of books for future inclusion in the series.

Published titles in the series

Environmental Risk Harmonization:
Federal and State Approaches to Environmental Hazards in the USA
Edited by Michael A. Kamrin (ISBN 0 471 972657)

Handbook of Soil Invertebrate Toxicity Tests
Edited by Hans Løkke and Cornelius A. M. van Gestel (ISBN 0 471 971030)

Pollution Risk Assessment and Management:
A Structured Approach
Edited by Peter E. T. Douben (ISBN 0 471 972975)

Forthcoming titles in the series

Behaviour in Ecotoxicology
Edited by Giacomo Dell'Omo (ISBN 0 471 968528)

Ecotoxicology of Wild Mammals
Edited by Richard Shore and Barnett Rattner (ISBN 0 471 974293)

Demography in Ecotoxicology
Edited by Jan E. Kammenga and Ryszard Laskowski (ISBN 0 471 490024)

Bioremediation and Land Reclamation:
Tools to Measure Success or Failure
Edited by Geoffrey Sunahara *et al.* (ISBN 0 471 986690)

Predictive Ecotoxicology
Mark Crane (ISBN 0 471 984043)

Statistics in Ecotoxicology

Edited by

Tim Sparks
Institute of Terrestrial Ecology,
Cambridgeshire,
UK

JOHN WILEY & SONS, LTD

Chichester · New York · Weinheim · Brisbane · Singapore · Toronto

Other Wiley Editorial Offices

John Wiley & Sons Inc., 605 Third Avenue,
New York, NY 10158-0012, USA

WILEY-VCH Verlag GmbH, Pappelallee 3,
D-69469 Weinheim, Germany

Jacaranda Wiley Ltd, 33 Park Road, Milton,
Queensland 4064, Australia

John Wiley & Sons (Asia) Pte Ltd, 2 Clementi Loop #02-01,
Jin Xing Distripark, Singapore 129809

John Wiley & Sons (Canada) Ltd, 22 Worcester Road,
Rexdale, Ontario M9W 1L1, Canada

Library of Congress Cataloging-in-Publication Data
Statistics in ecotoxicology / edited by Tim Sparks.
      p.   cm.   (Ecological & environmental toxicology series)
   Includes bibiliographical references.
   ISBN 0-471-96851-X (hc. : alk. paper)
   ISBN 0-471-97299-1 (pbk. : alk. paper)
   1. Pollution--Environmental aspects--Statistical methods.
   I. Sparks, Tim.   II. Series.
   QH545.A1S694   2000
   577.27'07'27—dc21   99–42708
                                                       CIP

British Library Cataloguing in Publication Data
A catalogue record for this book is available from the British Library

ISBN 0-471-96851-X (CL)
ISBN 0-471-97299-1 (PR)

Typeset in 10/12pt Garamond from the author's disks by Footnote Graphics, Warminster, Wilts.
Printed and bound in Great Britain by Antony Rowe Ltd, Chippenham, Wiltshire
This book is printed on acid-free paper responsibly manufactured from sustainable forestry,
in which at least two trees are planted for each one used for paper production.

"Have we statisticians given sufficient attention to guiding the scientist faced with the perennial difficulty of choosing, from among the vast array of statistical techniques now available, that which is appropriate to his or her current research?"

D. J. Finney

Reproduced by permission of Taylor & Francis Ltd,
from Journal of Applied Statistics, 22, 304.
http://www.tandf.co.uk/journals/.htm

Contents

List of contributors

Christine Anderson-Cook
Department of Statistics, Virginia Tech, Blacksburg, VA 24061-0439, USA

Ralph Clarke
NERC Institute of Terrestrial Ecology, Furzebrook Research Centre, Wareham, Dorset BH20 5AS, UK

Loveday L. Conquest
College of Ocean & Fishery Sciences, University of Washington, Seattle, WA 98195-7980, USA

Mark Crane
School of Biological Sciences, Royal Holloway, University of London, Egham, Surrey TW20 0EX, UK

Lois Dale
NERC Institute of Terrestrial Ecology, Monks Wood, Abbots Ripton, Huntingdon, Cambridgeshire PE17 2LS, UK

Don French
NERC Institute of Terrestrial Ecology, Banchory Research Station, Hill of Brathens, Glassel, Banchory, Kincardineshire AB31 4BY, UK

Dick Gadsden
School of Computing and Management Sciences, Sheffield Hallam University, Pond Street, Harmer Building, Sheffield S1 1WB, UK

Albania Grosso
WS Atkins Environment, Woodcote Grove, Ashley Road, Epsom, Surrey KT18 5BW, UK

Colin Janssen
Laboratory for Biological Research in Aquatic Pollution, University of Ghent, Plateaustraat 22, 9000 Ghent, Belgium

David Lindley
NERC Institute of Terrestrial Ecology, Merlewood Research Station, Grange-over-Sands, Cumbria LA11 6JU, UK

Reinhard Meister
Technische Fachhochschule Berlin, FB II Mathematik/Physik/Chemie, Luxemburger Str. 10, D-13353 Berlin, Germany

Ian Newton
NERC Institute of Terrestrial Ecology, Monks Wood, Abbots Ripton, Huntingdon, Cambridgeshire PE17 2LS, UK

Dan Osborn
NERC Institute of Terrestrial Ecology, Monks Wood, Abbots Ripton, Huntingdon, Cambridgeshire PE17 2LS, UK

Peter Rothery
NERC Institute of Terrestrial Ecology, Monks Wood, Abbots Ripton, Huntingdon, Cambridgeshire PE17 2LS, UK

Andy Scott
NERC Institute of Terrestrial Ecology, Merlewood Research Centre, Grange-over-Sands, Cumbria LA11 6JU, UK

Eric P. Smith
Department of Statistics, Virginia Tech, Blacksburg, VA 24061-0439, USA

Tim Sparks
NERC Institute of Terrestrial Ecology, Monks Wood, Abbots Ripton, Huntingdon, Cambridgeshire PE17 2LS, UK

A. J. Underwood
Centre for Research on Ecological Impacts of Coastal Cities, Marine Ecology Laboratories A11, University of Sydney, NSW 2006, Australia

Paul J. Van den Brink
DLO Winand Staring Centre, dept. Fate and Effect Pesticides, PO Box 125, 6700 AC Wageningen, The Netherlands

Foreword

The *Ecological & Environmental Toxicology Series* published by John Wiley & Sons aims to address new topics and perceived insufficiencies in the current literature. This new book Statistics in Ecotoxicology edited by Tim Sparks was long in gestation but is stronger for this fact. It is the first book within the series to be considered a textbook and offers insight into an area where many fear to tread; that of numbers, their interpretation and significance. The book is not intended to be an all encompassing book on statistics that may have relevance for ecotoxicologists, but rather is intended to provide useful statistical advice and information for the ecotoxicologist. This book will be useful for every scientist involved in the manipulation of, or required to interpret and understand their data. Although aimed as an introductory text it is destined to become one of those books in which people will find solace, rather like the blanket comforter of a child.

We hope that this book will meet the needs of those that have waited patiently for its publication. We will certainly know what to reach for when next troubled by that enigmatic data point.

Jason M. Weeks
Sheila O'Hare
Barnett Rattner

Preface

It used to be the case, and probably still is, that students of agriculture received a thorough grounding in field experimentation. That thoroughness does not extend, in my experience, to many of the other biological sciences. There may be many reasons for this; statistics viewed as a peripheral subject for future researchers, statistics training provided by non-statisticians, statistics training provided by theoretical statisticians(!), lack of numeracy being more sociably acceptable than a lack of literacy. Whatever the reason, a thorough understanding of the concepts of statistical methodology is essential for a scientific researcher to function effectively. This does not mean that we have to create statisticians from every practising biologist, but they must understand the potential of, and limitations of, statistical analysis. To misquote my late father, statistical methodology has made great strides in recent decades, some of them forwards. Rapidly improving statistical software enables us to be much more adventurous in our analysis, we need to ensure it is being used wisely and correctly.

This book attempts to provide an introduction to statistics in ecotoxicology (that strange hybrid between ecology and toxicology), explaining the concepts by illustration with a minimum of mathematical jargon. If this is successful then the contributors merit praise, if this fails then the blame rests solely on my shoulders. Introductory chapters deal with basic concepts and with exploratory data analysis. The next five chapters are then dedicated to field experimentation, lab experimentation, regression methodology, multivariate methods and monitoring. Finally, three case study chapters deal with terrestrial, freshwater and marine topics.

This book has had a long gestation, and I extend apologies to the various contributors and to Janie Curtis at Wiley for this. I would like to dedicate this book to my neglected family: Jill, Joe and Alex.

ACKNOWLEDGEMENT

I am grateful, as ever, for the wise counsel of my colleague Peter Rothery.

Tim Sparks
Monks Wood
May 1999

1

Basic Concepts

PETER ROTHERY

Institute of Terrestrial Ecology, Monks Wood, UK

1.1 INTRODUCTION

This chapter discusses some basic statistical concepts as background for the methods of analysis presented in this book. The need for statistical analysis in ecotoxicology, and biology in general, arises from the variability which is an essential characteristic of individuals and biological populations. In particular, we emphasize random variation and its implications for studying biological populations using samples of observations from some defined statistical population. We categorize the main types of data and describe some ways of summarizing data which apply to populations and samples. We introduce the basic ideas of probability for describing the outcome of random processes in which there is a chance element, and discuss the properties of some important probability distributions for describing random variation in both continuous and discrete measurements. We then apply these ideas to the general problems of estimation and testing a statistical hypothesis. The concepts and methods developed in this chapter apply to a wide range of ecotoxicological studies. However, to emphasize differences of interpretation we make the important distinction between experiments, in which material is deliberately manipulated to establish causal links, and surveys which are essentially observational and restricted to the description of pattern and correlation.

The emphasis of this chapter is on concepts rather than methods, although some particular methods are used for illustration. The following chapter describes in detail a range of techniques for exploring data, including graphical methods, tables and summary measures.

Statistics in Ecotoxicology. Edited by T. Sparks. © 2000 John Wiley & Sons Ltd

1.2   POPULATIONS, VARIABILITY AND SAMPLES

1.2.1   BIOLOGICAL AND STATISTICAL POPULATONS

In ecotoxicological studies the data are generally observations or measurements made on individuals which are a sample from some biological population of interest that has been subjected to elevated levels of a pollutant. For example, the measurements of mercury, cadmium and selenium concentrations in the kidney and liver in a sample of 13 adult skuas taken from a population of ringed birds at Foula, Shetland in 1976 (Appendix 1). The biological population refers to a group of individuals of the same species inhabiting a given area during some time. The techniques described in this book, however, apply to a more general population type, namely a statistical population which refers to a collection of sampling units from which the samples are drawn. The statistical population could be the measurements on different individuals in some biological population, as in the skua data. It could also be a population of measurements of an individual sampled on different occasions. In a study to estimate the density of earthworms in a field from core counts, the statistical population can be visualized as a spatial distribution of cores. A more abstract example occurs in the experiment to investigate the effect of different doses of copper on the behaviour of beetles (Appendix 2). For each dose, we can think of a statistical population of responses that would have been obtained had that dose been applied to all of the 79 individuals, or to some wider population from which they were selected. The statistical population may be finite, infinite, real or conceptual. A basic problem in statistics is to estimate population values of a statistical population from a sample of observations.

1.2.2   VARIABILITY IN BIOLOGICAL POPULATIONS

Variability is a conspicuous feature of biological organisms, populations and communities. In many situations some of the variation between individuals can be systematically related to factors such as the sex or age, or some measurable aspect of its environment such as food supply or the dose level in a bioassay. Often, however, there is haphazard variation which cannot be accounted for, and there is little choice but to regard the variability as inherently random. In many situations random variability is regarded as a nuisance because it tends to obscure differences between populations. On the other hand random variability may be of primary interest – for example in a study of the individual variation in tolerance to some toxin. Random variation in biological systems can arise in many ways, but we can identify three broad types, namely natural variability, measurement error and sampling error.

1.2.2.1 Natural variability

This is random variation which is an inherent feature of the individual or population. Natural variability arises from internal sources such as genetic variation and physiological fluctuations. Genetic recombination is one of the most fundamental of all biological processes, and this appears to involve randomness. Organisms are also affected by external sources of variation from changes in their biotic and physical environment which have a random component. For example, the effect of weather which can vary in an unpredictable way.

1.2.2.2 Measurement error

The measured value of a biological variable for an individual will usually contain some measurement error. This error is distinct from natural variability, but it is a part of the observation. The measurement error may contain a systematic error or bias, a random component, or both. Ideally, measurement errors should be made small, or at least small relative to differences due to natural variation. Often this is achieved by refining the measuring instrument and technique. One way to reduce random measurement error is to use the average of several independent measurements; but this would not reduce any bias.

1.2.2.3 Sampling error

Most studies of biological populations are based on a sample of individuals, as it is not generally possible to carry out a total census. Usually, the purpose of the sample is to infer something about the population from which it was selected. Because of the essential variability of biological populations, information provided by a single sample is incomplete and varies between samples taken from the same population. Sampling error refers to the uncertainty in the inferences about the population which are drawn from a sample. Section 1.6 shows how the uncertainty can be measured using for illustration the problem of estimating the population mean. First, we introduce some basic methods of sampling and related concepts.

1.2.3 RANDOM SAMPLING, RANDOMIZATON AND INDEPENDENCE

Most statistical methods assume that there is a random element in the selection of the units to be included in the sample. The basic procedure is simple random sampling which gives each possible sample of the same size an equal chance of selection. This is like a fair lottery or dealing a hand of cards from a well-shuffled pack. Simple random sampling is important for two reasons. First, it avoids selection bias, i.e. no particular individual is any more or less likely than any other to be included in the sample. Second, it provides estimates of population values with known properties, and a method for measuring the errors of the estimates. These properties refer to what happens in the hypothetical long-run of a very large number of repeated samples from the same population; there is no guarantee that a particular sample will produce a good estimate. A simple random sample could be drawn from a list of all the units in the population, but in practice this may not be feasible, e.g. when sampling a starling roost. Biases can occur when the chance of selection of an individual is related to the measurement of interest, e.g. birds with low levels of pollutant may be more difficult to catch.

Randomization in experiments is an objective procedure for allocating treatments at random to the experimental units. This safeguards against bias and ensures that any differences may be reasonably attributed to the effects of treatments. The equivalent of simple random sampling is a completely randomized design, in which all possible allocations of the treatments are equally likely. An example is the experiment to study the effect of different doses of copper on the behaviour of beetles.

In simple random sampling and randomization no attempt is made to allow for any heterogeneity in the population. Often, however, there may be systematic differences in the response of different groups of individuals, e.g. in a mixed population, females may contain higher levels of a pollutant than males. In these cases, it may be more effective to stratify the population and then to sample at random within strata. These aspects of design are discussed in later chapters.

A key property of simple random sampling and randomization is that the observations on the units included in the sample are independent of each other. Lack of independence can occur when the units are grouped in some way. For example, suppose that in the skua data the sample of 12 skua eggs had been obtained by taking two eggs from each of six nests rather than from 12 different nests. Then, the 12 observations are not independent because pollutant levels in eggs from the same nest are more likely to be similar than levels in eggs from different nests. Essentially there are six independent observations. A similar situation arises in experiments where treatments are allocated to animal litters. Pseudoreplication occurs when non-independent observations are treated as if they were independent and this usually leads to spurious accuracy (Chapter 3).

Lack of independence also arises when the data are a series in time, e.g. levels of contaminants measured at monthly intervals in a catchment area. Observations which are close together in time are usually more similar than those which are further apart, an effect which is referred to as autocorrelation or serial correlation. The statistical analysis of time-series data is a specialized topic which is beyond the scope of this book.

1.3 MEASUREMENT SCALES AND TYPES OF DATA

Statistical techniques depend on the type of data available. The choice of method should reflect the scale of measurement and it is useful to distinguish between nominal, ordinal, interval and ratio scales.

1.3.1 NOMINAL SCALE (CATEGORICAL DATA)

Data on a nominal scale arise when individuals are grouped into categories or groups. The categories must be mutually exclusive, i.e. each individual belongs to only one group. For example, gender (male/female); blood group (A, B, O, AB); state (live/dead); stage (egg, larva, pupa); species (sparrowhawk, barn owl, heron). The term nominal indicates that there is no natural ordering of the categories. Binary data are a special case of just two possible outcomes. Categorical data are often presented as the proportions of individuals belonging to each group. This is useful for comparison of samples in which the total number of individuals is different.

1.3.2 ORDINAL SCALE (RANK DATA)

An ordinal scale is a nominal scale in which the classes can be placed in order or ranked. For example, in a vegetation survey cover in a quadrat may be allocated into one of six classes: 1 (0%); 2 (1%–5%); 3 (6%–10%); 4 (11%–20%); 5 (21%–50%); 6 (51%–100%). In behavioural studies, relative dominance is sometimes measured using an individual's rank in an observed pecking order. Note that differences between ranks are not usually meaningful, e.g. the difference between rank 2 and rank 1 is not the same as that between rank 12 and rank 11.

1.3.3 INTERVAL AND RATIO SCALES

Observations on an interval scale can be ordered in terms of the measurements (as in rank data) and the intervals between them. A difference of one unit of measurement is the same at all points along the scale. Examples are dates, distances, temperatures and weights. A ratio scale is an interval scale which includes an absolute zero, e.g. weight or pollutant level. Ratios of measurements are meaningful quantities irrespective of the units of measurement. For example, in the skua eggs (Appendix 1) the highest level of PCB (36 ppm) was six times the lowest level (6 ppm), a sixfold range.

1.3.3.1 Continuous measurements and count data

For interval data a useful distinction is between continuous measurements and counts. Continuous measurements can theoretically take any value over some range, e.g. temperature, weight, although in practice the possible values are limited by the precision of the measuring instrument. Count data are necessarily restricted to a range of the whole numbers (integers) 0,1,2, . . . etc. although not all values within the range need occur. Examples are the number of leaves on a plant, the number of earthworms in a core sample, the number of survivors at a given dose in a bioassay, the latter being derived from a number of individual binary responses. Section 1.5 presents some models for describing random variation in continuous and discrete variables. Chapter 5 discusses some associated methods for analysing statistical relationships for different types of response data including linear regression (continuous) and logistic regression (binary) and touches on log-linear models or Poisson regression (counts).

1.4 SUMMARY MEASURES FOR SAMPLES AND POPULATIONS

In dealing with samples of biological data we need methods for summarizing and presenting data. This is necessary for large amounts of data where interesting features may not be clear from scanning tables of figures. Data summaries are also useful in a preliminary analysis to compare samples taken from different populations. Two important aspects of a sample are the location and the spread. The most commonly used measures of location are the mean, median and the mode. Spread is most often measured by the variance or the standard deviation. These measures are discussed below using for illustration the measurements of PCB in the sample of 12 skua eggs. Measures of location and spread can also be defined for the statistical populations from which samples are drawn. A key idea in statistical inference is to use sample values (statistics) to estimate the corresponding population values (parameters). This section deals with the basic numerical summaries. Chapter 2 extends the discussion to graphical summaries and presentation, and other numerical measures.

1.4.1 MEASURES OF LOCATION

1.4.1.1 Mean

This usually refers to the arithmetic mean, i.e. the average value obtained by adding up the figures and dividing by the number of observations. The mean level of PCB (ppm) in the sample of 12 skua eggs is

$$\frac{1}{12}(19+13+10+33+10+36+19+30+13+6+33+21) = 20.25 \, \text{ppm} \quad (1.1)$$

For a sample of size n with values denoted by $x_1, x_2, \ldots, x_n$ the mean is often denoted by $\bar{x}$ (pronounced 'x bar') where

$$\bar{x} = \frac{1}{n}(x_1 + x_2 + x_3 + \ldots + x_n) = \frac{1}{n}\sum_{i=1}^{n} x_i \qquad (1.2)$$

The symbol Σ is the Greek capital letter sigma and is often used in statistics as a shorthand notation to denote summation.

If there are two samples of size n_1 and n_2 with means $\bar{x}_1$ and $\bar{x}_2$, then the overall mean is calculated as the total for the two samples divided by the total number of values ($n = n_1 + n_2$). This can be calculated as $(n_1\bar{x}_1 + n_2\bar{x}_2)/n$, i.e. a weighted mean of the two samples means with weights proportional to sample size.

1.4.1.2 Median

The median is the middle value when the data are arranged in order of magnitude. When there is an even number of observations the median is the average of the two values in the middle. For the 12 skua PCB measurements the ordered values are

$$6, 10, 10, 13, 13, [19, 19], 21, 30, 33, 33, 36 \qquad (1.3)$$

with the median as the average of the sixth and seventh values, i.e. $(19 + 19)/2 = 19$ ppm.

For a large population of continuous measurements, 50% of values are below the median and 50% of values are above it. For discrete data there may be fewer than 50% of values above or below the median because of the possibility of values actually equal to the median.

The median is less affected by extreme data values than the mean. For example, if the largest concentration of PCB in the skua was 60 ppm instead of 36 ppm, the mean would increase from 20.25 to 22.25 ppm, whereas the median would remain at 19 ppm.

1.4.1.3 Mode

The mode is the most frequently occurring value. For the skua sample there are three modes, i.e. 13, 19 and 33 ppm; these values each occur twice. This illustrates a limitation of the mode as a measure of location in a sample: modes may occur by chance because of random sampling, without there being any biological significance. Modes are not widely used in the statistical analysis of biological data. An exception is the analysis of data from populations which contain a mixture of populations each with a different mean, e.g. lengths of individuals from different cohorts or age classes.

1.4.2 MEASURES OF SPREAD

1.4.2.1 Variance and standard deviation

Variance and standard deviation both measure the extent to which the observations spread out from the mean. The variance of a statistical population is defined as the mean squared deviation from the population mean. The standard deviation is the square root of the variance. The two measures are directly related, but both are important for statistical analysis. The sample mean squared deviation is the sum of the squared deviations from the sample mean divided by the sample size (n). In practice, however, we use the sample variance calculated as the sum of the squared deviations from the sample mean divided by $n - 1$. We divide by $n-1$ rather than by n because the sample mean squared deviation tends to underestimate the population variance. For the PCB levels in the sample of 12 skua eggs (sample mean = 20.25 ppm) the sample variance is equal to

$$\frac{1}{11}[(19-20.25)^2 + (13-20.25)^2 + \ldots + (21-20.25)^2] = 108.20 \text{ ppm}^2 \quad (1.4)$$

The sample standard deviation is then $\sqrt{108.20} = 10.40$ ppm. Note that the standard deviation is in the same units as the data.

The general formula for the sample variance of a set of n measurements denoted by $x_1, x_2, \ldots, x_n$ is given by

$$s^2 = \frac{1}{(n-1)} \sum_{i=1}^{n} (x_i - \bar{x})^2 \quad (1.5)$$

The corresponding standard deviation is denoted by s.

For measurements which cannot take negative values (e.g. levels of PCB) the coefficient of variation is defined as the ratio of the standard deviation to the mean. This is a dimensionless quantity which measures the spread relative to the mean. For the skua data the sample coefficient of variation ($CV = s/\bar{x}$) is calculated as $10.40/20.25 = 0.51$, or 51%. The variance and standard deviation are the most widely used measures of spread. This is partly because they arise in the specification of the so-called normal distribution, a curve for describing the frequency of values in a population of continuous measurements and which is used as a basis for a wide range of statistical methods. It is important to stress the distinction between the standard deviation and the so-called standard error of the mean, often shortened to standard error (se). The standard deviation measures the spread of the individual observations, whereas the standard error of the mean refers to the spread of the sample mean calculated from repeated random samples from the population. For a sample of size n with standard deviation s, the standard error of the mean is calculated as $s/\sqrt{n}$.

1.4.2.2 Mean absolute deviation, range and inter-quartile range

In some situations other measures of spread can be useful. The mean absolute deviation (MAD) is the average of the absolute deviations from the mean. Using absolute deviations (rather than squared values as used for the variance) gives less weight to the more extreme observations. This could be an advantage if there were occasional rogue data values or outliers which were not representative of the population of interest. The range measures spread as the difference between the two extreme values. A limitation of the range is that it is likely to increase as the number of observations in the sample increases. This problem can be avoided by using the inter-quartile range which is the difference between the upper and lower quartiles (i.e. the values separating the upper 25% and lower 25% of the observations).

1.5 PROBABILITY AND PROBABILITY DISTRIBUTIONS FOR DESCRIBING RANDOM VARIATION

Random variation as an inherent part of our observations and measurements on biological populations and processes. In section 1.4 we presented some summary measures of location and spread of the observations in a sample of data, or in the statistical population from which the sample was selected. In this section we discuss probability as a basis for describing random variation and present some basic models.

1.5.1 PROBABILITY, RANDOM VARIABLES AND PROBABILITY DISTRIBUTIONS

Probability is a part of everyday experience, e.g. lotteries, games of chance, predictions of uncertain events such as the weather and assessing risks of disease. There are several ways to define probability, but we shall adopt a so-called Frequentist view which regards probability as a long-run relative frequency. This approach can be illustrated by the results of an experiment in which a coin was tossed 300 times and the sequence of events 'heads' (H) or 'tails' (T) was recorded. Figure 1.1. summarizes the data as the accumulated number of 'heads' expressed as a proportion of the number of tosses, e.g. the first five events were HTTTH with cumulative proportions of heads equal to 1/1, 1/2, 1/3, 1/4, 2/5. As the number of tosses increases the proportion of 'heads' tends to settle down and is close to the value of 0.5. This motivates a definition of probability as a long-run relative frequency or proportion in a very large number of identical repetitions of a random trial. In practice, we cannot observe the hypothetical long-run proportion, but refer to the probability of a particular event occurring in a given trial, e.g. the probability of the event 'heads' is 0.5.

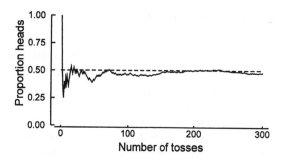

Figure 1.1 Illustration of frequency definition of probability, showing how cumulative proportion of 'heads' in successive tosses of a coin approaches a stable value 0.5 in the long-run.

Similarly, an individual selected at random from a population with 50 : 50 sex ratio would result in a female with probability 0.5.

The number of possible outcomes in the trial is not restricted to two. In a single throw of a die there are six possible outcomes (1, 2, 3, 4, 5, 6), each with probability 1/6 (assuming that the die is not loaded). A random sample of five individuals from a population of males and females could contain (0, 1, 2, 3, 4, 5) males: the associated probabilities are obtained from the so-called binomial distribution (section 1.5.2).

1.5.1.1 Random variables and probability distributions

A random variable is one which is subject to chance variation. For example, in a single toss of a coin the events 'heads' and 'tails' can be represented as the random variable R which takes the value 1 for a 'head' and 0 for a 'tail'. The probability of a 'head' is written as $P(R = 1) = 0.5$. In this case, the random variable is discrete, being restricted to two possible values. Random variables can also be continuous measurements, for example the level of PCB in an egg from a randomly selected nest in a breeding population of skuas.

The probability distribution of a discrete random variable specifies the possible values that the random variable can take, together with a list of associated probabilities. For a continuous random variable this approach does not work because the infinitely many possible values in a given range imply that the probability of a particular value is zero. To solve this problem we represent the values of the probability by a smooth curve called the probability density function (p.d.f.). The probability that a continuous random variable lies between two values is given by the area under the curve between the two values.

The cumulative distribution function (c.d.f.) gives the probability that the random variable takes a value less than or equal to a specified value. For example, if R is a discrete random variable which can take only positive values, the c.d.f. is given by

$$F(r) = P(R = 0) + P(R = 1) + \ldots + P(R = r) \qquad (1.6)$$

For the continuous case, $F(x)$ is the area under the probability density function up to and including the value x.

1.5.1.2 Mean and variance of a probability distribution

In section 1.4 we discussed the mean and variance as measures of location and spread in a set of observations. Here we show how these measures apply to random variables by defining them for a probability distribution. We present the results for discrete distributions but they apply in an analogous way to continuous distributions.

The mean can be thought of as the centre of gravity of the distribution. For the discrete case, the mean is calculated as a weighted average of the possible values with weights equal to the probabilities, i.e. for a random variable R,

$$\text{Mean} = \sum_{\text{all } r} rP(R = r) \qquad (1.7)$$

Another term for the mean is expectation or expected value, written as $E[R]$ and often denoted by μ (the Greek letter 'mu').

Variance is defined as the mean squared deviation about the mean, i.e. a weighted average of the squared deviations with weights equal to the probabilities and given by

$$\text{Variance} = \sum_{\text{all } r} (r - \mu)^2 P(R = r) \qquad (1.8)$$

This is often written as $\text{Var}[R]$ and denoted by σ^2. The standard deviation of the distribution is the square root of the variance, i.e. σ (the Greek letter 'sigma').

If a random variable is multiplied by a constant k, then the mean and standard deviation are also multiplied by k, and the variance is multiplied by k^2. Means and variances also have useful properties. First, the mean of a sum of random variables is equal to the sum of the individual means. Second, the variance of a sum of random variables is equal to the sum of the variances, provided that they are independent, i.e. they do not tend to fluctuate together.

1.5.1.3 Mean and variance of the sample mean

An important application of the above results relates to the properties of the mean of a random sample from a large population. We regard the n observations as independent random variables each with the same mean (μ) and variance (σ^2) because they are drawn at random from the same population. The mean of the sample total is equal to $n\mu$ and the variance is $n\sigma^2$. So, dividing by n, the mean of the sample mean is μ and the variance is σ^2/n, with standard deviation $\sigma/\sqrt{n}$. In repeated random samples from the same population the spread of the sample mean (measured by its standard deviation) increases in proportion to the population standard deviation and declines inversely with the square root of the sample size. See section 1.5.7 for a numerical illustration of this result, and section 1.6 for further discussion on how the result is used to measure the error in estimating a population mean.

In the following sections we summarize the properties of some of the more important discrete and continuous distributions.

1.5.2   THE BINOMIAL DISTRIBUTION

There are many situations in ecotoxicology where the observation on an individual takes one of two possible values, e.g. an animal may die from a dose of some toxin, or it may survive. Such observations are referred to as binary and often coded as 1 for a 'success' and 0 for a 'failure'. In a bioassay, there may be several animals receiving the same dose, and a total (or proportion) of survivors is derived. Individuals vary in their response so the number of survivors is a random variable. The binomial distribution is a probability distribution for the total number of successes in a series of trials with only two outcomes, where: (a) the probability of success is constant from trial to trial; (b) the trials are independent of each other.

To illustrate the binomial distribution consider a coin-tossing experiment involving three tosses of the coin, and the calculation of the probability distribution of the total number of 'heads' (R). There are eight possible outcomes: TTT; TTH; THT; HTT; THH; HTH; HHT; HHH, where T and H represent the outcome 'tails' and 'heads' respectively. The three basic events are independent and occur with probability 1/2, so the probability of a particular sequence is $1/2 \times 1/2 \times 1/2 = 1/8$. The probability of obtaining a given value of R is obtained by summing the probabilities of the sequences giving rise to that value. For example, when $R = 2$ the sequences are THH, HTH and HHT, so that $R = 2$ with probability $1/8 + 1/8 + 1/8 = 3/8$. The probability distribution of R is then calculated as: $P(R = 0) = 1/8$; $P(R = 1) = 3/8$; $P(R = 2) = 3/8$; $P(R = 3) = 1/8$.

Now consider the more general case of the number of 'successes' R in n independent trials each with probability of 'success' p. The probability of a particular sequence with r 'successes' and $n - r$ 'failures' is equal to $(1 - p)^{n-r}$. The number of sequences with $R = r$ is the number of different ways of selecting the r objects from n and written as ^nC_r. The general formula for the binomial distribution is then

$$P(R = r) = {}^nC_r\, p^r\, (1 - p)^{(n-r)} \quad \text{for } r = 0, 1, 2, \ldots, n \tag{1.9}$$

The number of combinations of r from n is calculated using the formula

$$^nC_r = \frac{n!}{r!(n-r)!} \tag{1.10}$$

where $n!$ (read 'n factorial') is the product $n \times (n-1) \times (n-2) \ldots 3 \times 2 \times 1$. For the coin-tossing example,

$$^3C_2 = \frac{3!}{2!1!} = \frac{3 \times 2 \times 1}{2 \times 1} = 3 \tag{1.11}$$

In practice, statistical packages usually have procedures for calculating the binomial probabilities.

Figure 1.2 shows the binomial distribution for various values of n and p, and illustrates the following general properties. The shape of the distribution depends on the value of p. For p equal to 0.5, the distribution is symmetric for all values of n. For values of $p < 0.5$, the higher probabilities occur at lower values of r, i.e. the distribution is positively skewed or skewed to the right. For values of $p > 0.5$, the distribution is negatively skewed. As n increases the distribution becomes more symmetrical.

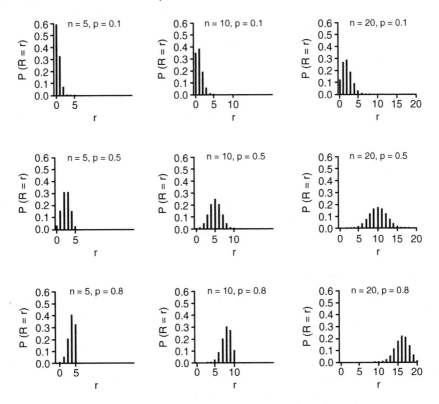

Figure 1.2 Binomial distributions for various values of n and p.

The mean of the binomial distribution is np, i.e. the expected number of 'successes' in n trials. The variance is $np(1 - p)$, which increases to a maximum value of $n/4$ when p is equal to 0.5.

1.5.2.1 Dose–response curves and logistic regression

An important application of the binomial distribution arises in toxicity studies where groups of individuals are subject to different doses of some toxin. For a group of n_x individuals receiving a particular dose x, the number of individuals surviving is assumed to follow a binomial distribution with mean $n_x p_x$, where the probability of survival p_x is related to the level of the dose, usually by a sigmoidal or S-shaped curve. Chapter 5 discusses the logistic regression model which is often used to describe this type of response.

1.5.3 THE POISSON DISTRIBUTION

In the previous section we described the binomial distribution for the number of successes in a fixed number of independent trials each with the same probability of success. In many situations, however, there is no obvious upper limit on the count, e.g. the number of nematodes in a core sample or the number of pollution incidents occurring in a given time period. The Poisson probability distribution is given by

$$P(R = r) = \frac{\mu^r}{r!} \exp(-\mu) \tag{1.12}$$

where $r = 0, 1, 2 \ldots.$ etc. and $r! = r \times (r - 1) \times (r - 2) \times \ldots 2 \times 1$. The parameter μ is the mean of the distribution. An important property is that the variance is equal to the mean, i.e. $\mathrm{Var}[R] = \mu$.

Figure 1.3 shows some Poisson distributions with different means. For smaller values of the mean the distribution is positively skewed, but as the mean increases the distribution becomes more symmetric.

1.5.3.1 The Poisson process

In a Poisson process events occur completely at random at a uniform rate in time (or space), i.e. the probability of an event in any small time interval is constant, and events occur independently of each other. To determine the distribution of the number of events in time T, we split the interval up into a large number of small time intervals of length T/n, in which the probability of an event is $p = \lambda T/n$. The number of events R then follows a binomial distribution with mean λT, and which approximates to a Poisson distribution (section 1.5.3.2).

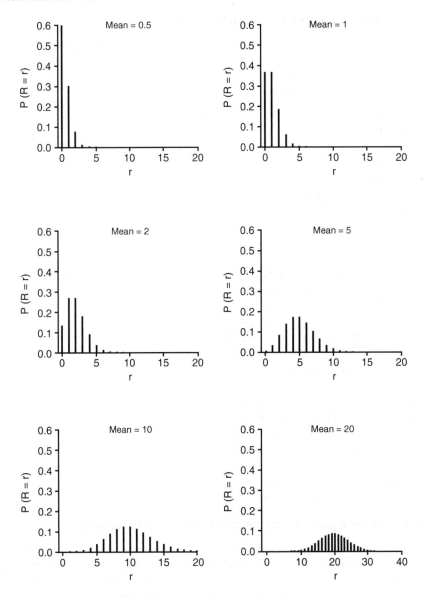

Figure 1.3 Poisson distributions with different means.

1.5.3.2 Poisson approximation to binomial

This section illustrates numerically how a Poisson distribution approximates to a binomial with the same mean. MINITAB commands to calculate probabilities for a binomial distribution with $n = 40$, $p = 0.05$, and a Poisson distribution with mean $\mu = 2$ are as follows:

```
MTB > set 'r'
DATA> 0:40
DATA> end
MTB > pdf 'r' 'bin_r';
SUBC> binomial 40 0.05.
MTB > pdf 'r' 'poi_r';
SUBC> poisson 2.
PRINT 'r' 'bin_r' 'poi_r'

Row   r    bin_r        poi_r
  1   0    0.128512     0.135335
  2   1    0.270552     0.270671
  3   2    0.277672     0.270671
  4   3    0.185114     0.180447
  5   4    0.090122     0.090224
  6   5    0.034151     0.036089
  7   6    0.010485     0.012030
  8   7    0.002680     0.003437
  9   8    0.000582     0.000859
 10   9    0.000109     0.000191
 11  10    0.000018     0.000038
 .  .  .  .
Remaining values omitted
```

1.5.3.3 Overdispersion and the negative binomial distribution

A property of the Poisson distribution is that the variance is equal to the mean. In many situations, however, the variance of a count distribution is greater than the mean, i.e. there is overdispersion and the distribution is said to be over-dispersed. For example, the number of nematode worms in a core may be overdispersed because of spatial variation in density caused by heterogeneity in the soil. Many distributions have been proposed for overdispersed count data, but one which has been widely used is the negative binomial distribution. For a negative binomial distribution with mean μ, the variance is equal to $\mu + \mu^2/k$, where the parameter k is called the index of dispersion.

When the variance is less than the mean the count distribution is under-dispersed. An example, is the binomial distribution with variance equal to

$np(1 - p)$. Underdispersion can result from a more regular spacing of individuals caused by inhibition or competition.

1.5.4   THE NORMAL OR GAUSSIAN DISTRIBUTION

The normal or Gaussian distribution is widely used for describing variation in continuous measurements, such as human height. Also, the distribution is basic to statistics because of the remarkable result that the mean of a sample of independent measurements is approximately normally distributed, whatever the distribution of the individual measurements. This is the central limit theorem, which is demonstrated below.

The normal probability density function is a bell-shaped curve

$$\frac{1}{\sigma\sqrt{2\pi}}\exp\left[-\frac{1}{2}\left(\frac{x-\mu}{\sigma}\right)^2\right]$$ (1.13)

where exp stands for the exponential function, and where x can take any value in the range from minus infinity to plus infinity (i.e. $-\infty < x < \infty$). The constant $1/\sigma\sqrt{2\pi}$ makes the total area under the curve equal to 1.

Figure 1.4 shows some normal distributions for different values of μ and σ. The parameter μ determines the position of the mode where the probability density is largest; the distribution is symmetric so that μ is also the mean and the median; σ is the standard deviation and σ^2 is the variance (section 1.5.1).

The cumulative distribution function is a symmetric sigmoidal (or 'S'-shaped) curve with value at the mean (median and mode) equal to 0.5 (Figure 1.5), and a slope which increases as the standard deviation decreases.

An important property of any normal distribution is that the area under the curve between any two multiples of σ on either side of the mean does not depend on μ and σ. In particular, the probability between $\mu - \sigma$ and $\mu + \sigma$ is 0.683, and the probability between $\mu - 2\sigma$ and $\mu + 2\sigma$ is 0.954. In other words, if an observation is drawn at random from a normal distribution then the probability that it falls within two standard deviations from the mean is equal to 0.954. The range $\mu \pm 1.96\sigma$ includes 95% of the probability and is often used in practical applications.

1.5.4.1   The standardized normal distribution

The standardized normal distribution has mean zero and variance one. Every normal distribution can be reduced to standardized form by (a) moving the distribution, i.e. a change of location, and then (b) stretching or contracting the distribution, i.e. a change in the scale. This can be achieved by subtracting the mean, and then dividing by the standard deviation to form a standardized variable given by $Z = (X - \mu)/\sigma$. For the normal distribution, there is no simple

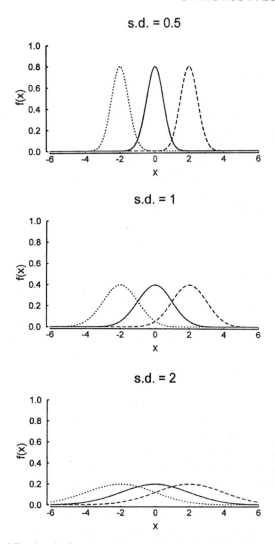

Figure 1.4 Probability density functions of normal distributions with different means $(-2, 0, 2)$ and standard deviations $(0.5, 1, 2)$.

formula for the probability that a value falls in a given range, and probabilities have to be calculated using numerical methods. However, any probability $P(X < b)$ can be reduced to $P(Z < (b - \mu)/\sigma)$, so that a single table of probabilities for values of the standardised normal distribution can be used. Note that for $a < b$,

$$P(a < X < b) = P(X < b) - P(X < a) \qquad (1.14)$$

i.e. a difference between two cumulative probabilities. There are statistical

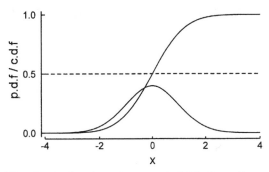

Figure 1.5 Probability density function of a normal distribution with mean zero and unit variance, together with the cumulative distribution function.

tables of cumulative probabilities for the standardized normal distribution, and most statistical packages have procedures for calculating them.

1.5.5 LOGNORMAL DISTRIBUTION

There are many situations in ecotoxicology where the measurements must take positive values and the variation is large relative to the mean. For example, the PCB level in the 12 skua eggs has mean $\bar{x} = 20.25$ ppm and standard deviation $s = 10.40$ ppm, i.e. a coefficient of variation $CV = 51\%$. Often, such measurements are more squashed up near zero and skewed towards larger positive values so that a normal distribution which is symmetric and allows negative values is therefore not appropriate. In these cases, the logarithm of the measurement is sometimes found to be approximately normally distributed. When the logarithm of a variable is normally distributed then the variable follows a lognormal distribution. If $\log_e X$ is normal with mean μ and variance σ^2, then X is lognormal with mean $\exp(\mu + \sigma^2/2)$ and variance $\exp(2\mu + \sigma^2)[\exp(\sigma^2) - 1)]$, with coefficient of variation $\sqrt{[\exp(\sigma^2) - 1]}$. Figure 1.6 illustrates the distribution for different values of σ.

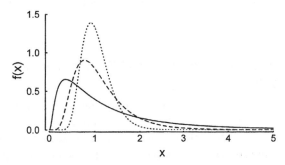

Figure 1.6 Lognormal distributions for $\log_e x$ normal with mean 0 and standard deviation 0.3 (dotted line), 0.5 (dashed line and 1.0 (solid line).

1.5.6   CHI-SQUARED, STUDENT'S t AND F DISTRIBUTIONS

The binomial, Poisson, normal and lognormal distributions have been widely applied to describe random variation in measurements on biological populations. Some other distributions are important because they are used for calculating probabilities associated with statistical methods, e.g. confidence intervals (section 1.6) and significance tests (section 1.7). Here we briefly mention the three key continuous distributions: chi-squared, Student's t and F. These are all related to the standardized normal distribution as follows.

1.5.6.1   Chi-squared distribution

The square of a standardized normal variable follows a chi-squared distribution with 1 degree of freedom, i.e. $\chi^2_1 = Z^2$ (where the symbol χ is the Greek letter 'chi'). More generally, the sum of squares of f independent standardized normal variables follows a chi-squared distribution with f degrees of freedom, i.e. $\chi^2_f = Z^2_1 + Z^2_2 + \ldots + Z^2_f$.

1.5.6.2   Student's t-distribution

The ratio of a standardized normal variable (Z) to the square root of an independent scaled chi-squared variable with f degrees of freedom χ^2_f/f follows Student's t-distribution with f degrees of freedom, i.e. $t_f = Z/\sqrt{(\chi^2_f/f)}$. When the degrees of freedom are large, Student's t-distribution approximates to a standardized normal distribution.

1.5.6.3 F-distribution

The ratio of a scaled chi-squared variable with f_1 degrees of freedom $\chi^2_{f_1}/f_1$ to an independent scaled chi-squared variable with f_2 degrees of freedom $\chi^2_{f_2}/f_2$ follows an F-distribution with f_1 and f_2 degrees of freedom, i.e. $F_{f_1,f_2} = (\chi^2_{f_1}/f_1)/(\chi^2_{f_2}/f_2)$.

1.5.6.4   Importance of chi-squared, t and F distributions for statistical analysis

One reason why these distributions are important stems from the statistical method of analysis of variance. This is a technique for partitioning the total variation in a set of observations into sums of squares measuring the contribution from different factors (Chapter 3). Ratios of sums of squares are used to assess whether the contribution of a particular factor is real rather than a chance effect.

1.5.7 SIMULATING RANDOM DATA

Many statistical packages have procedures for simulating random data from a range of probability distributions. These procedures are useful for studying properties of distributions and examining the performance of statistical methods. The following exercise illustrates the result that the mean of n independent observations from a normal distribution (μ, σ^2) follows a normal distribution (μ, σ^2/n). Means were obtained from 1000 sets of 5 observations from a normal distribution (10, 2^2) using the MINITAB package. Figure 1.7 summarizes the random data as a histogram and a normal curve has been superimposed for comparison.

1.5.8 CENTRAL LIMIT THEOREM

The central limit theorem says that the sum of a large number of independent random variables drawn from the same distribution is approximately normally distributed, whatever the original distribution. The approximation depends on the shape of the original distribution but becomes closer as the number of variables increases. Furthermore, the theorem also applies to the distribution of the mean, which explains the widespread use of the normal distribution in statistical analyses involving the sample mean. The central limit theorem may also explain why measurements which result from the combined effects of many factors are often approximately normally distributed. An example is human height which is determined in part by the actions of several genes.

The distribution of the mean can rapidly approach normality for some quite non-normal distributions, even for as few as 10 observations. The example below illustrates this for an exponential distribution which is markedly skewed from zero towards large positive values, very different to the normal distribution. The probability density function is $(1/\mu) \exp(-x/\mu)$, with mean

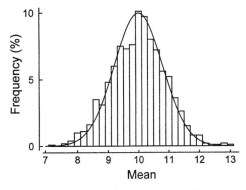

Figure 1.7 Illustration of sampling distribution of the mean of 5 observations from a normal distribution with mean 10 and variance 4.

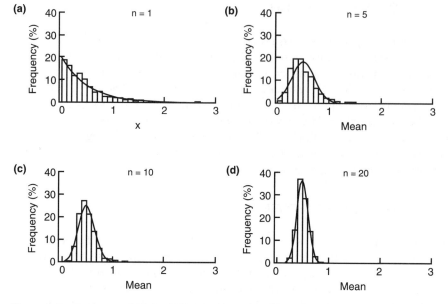

Figure 1.8 Illustration of the central limit theorem using samples of different sizes from an exponential distribution with mean 0.5. Histograms based on 1000 observations from each distribution. Curve in (a) is the probability density function of an exponential distribution with mean 0.5. Curves in (b)–(d) are normal distributions having the same mean ($0.5/n$) and variance ($0.5^2/n$) as the distribution of the sample mean.

$E[X] = \mu$ and variance $\text{Var}[X] = \mu^2$. The action of the central limit theorem was examined for a mean of 0.5 using samples of $n = 5$, 10 and 20. Figure 1.8 summarizes the simulated random data as a histogram, and in each case a normal distribution with appropriate mean μ and variance μ^2/n has been superimposed. As n increases the distribution approximates more closely to the normal.

1.6 ESTIMATING POPULATION PARAMETERS AND ASSESSING THE ACCURACY OF THE ESTIMATES

In most ecotoxicological studies measurements are made on a sample of individuals drawn from some population, and the aim is to use the sample to make inference about the population. For example, using the levels of PCB in a sample of 12 skua eggs to estimate the mean level of PCB in the breeding population. The problem arises because of the inherent variability in the measurements, so that different samples from the same population would generally give different estimates. In practice, we are dealing with a single sample, and what we would like to know is how close our estimate is to the unknown population value. The estimate will almost certainly be in error, but how can we measure the magnitude of the error in a useful way? In this section we show how errors of estimating population parameters can be measured by using the sample to calculate a range within which the true value of the parameter will lie with some prescribed probability (a confidence interval). For illustration we consider the problem of estimating a population mean using a random sample from a normal distribution. We then show how the method applies, at least approximately, to estimating the mean of any population. A confidence interval is a general method for measuring the error in a parameter estimate: an important application is in fitting models to data where confidence intervals are used to measure the error of the estimated parameters in the fitted models.

1.6.1 SAMPLING DISTRIBUTION OF THE SAMPLE MEAN

For a random sample the sample mean is the natural estimate of the population mean. However, different samples give different values for the mean, and the key to assessing the error in the sample mean is how it varies about the population mean from sample to sample. The sampling distribution of the sample mean describes how the sample mean varies in repeated random samples of the same size from the same population. The properties of the sample mean as an estimator of the population mean are usually described in terms of its sampling distribution. These are as follows:

1. The mean of the sampling distribution of the sample mean is equal to the population mean, i.e. random deviations from the mean tend to cancel each other out so that there is no systematic error or bias. We say that the sample mean is an unbiased estimator of the population mean. This property applies to the mean in a random sample from any population.
2. The variance of the sampling distribution of the sample mean is equal to the population variance divided by the sample size, i.e. σ^2/n. This says that the spread of the sampling distribution about the mean decreases with increasing sample size. In larger samples we therefore expect smaller errors. The magnitude of the error is usually measured by the standard error of the mean, the square root of its sampling variance, i.e. $\sigma/\sqrt{n}$. The result applies to the mean in a random sample from any infinitely large population. For a finite population of size N the variance is equal to $\sigma^2(1-f)/n$, where $f = n/N$ is called the sampling fraction.
3. The sampling distribution of the sample mean is normal. This result is exact when the distribution of the observations is normal, and applies approximately for other population distributions because of the central limit theorem (section 1.5.8).

The above results were illustrated in Figure 1.7 using samples of size 5 from a normal distribution. The properties can be used to calculate the likely error in using the sample mean to estimate the population mean. The method uses the result that in repeated sampling from any normal distribution, 95% of the deviations from the mean are within ±1.96 times the standard deviation (section 1.5). Applying this result to the sampling distribution of the sample mean shows that the sample mean will be within ±1.96 times the standard error in approximately 95% of cases. This result forms the basis of calculating a 95% confidence interval for the population mean described in the next section.

1.6.2 CALCULATING A CONFIDENCE INTERVAL FOR THE POPULATION MEAN

The definition and calculation of a confidence interval for the population mean rests on the key result from the previous section, i.e. in repeated random samples of size n from a normal distribution with mean μ and standard deviation σ, the sample mean $\bar{x}$ falls within the interval $\mu \pm 1.96\sigma/\sqrt{n}$ in 95% of cases. This result can be turned around as follows. In repeated random samples of size n from a normal distribution with mean μ and standard deviation σ, the random interval $\bar{x} \pm 1.96\sigma/\sqrt{n}$ will contain the population mean in 95% of cases.

In random samples from a normal distribution with standard deviation σ the interval calculated as $\bar{x} \pm 1.96\sigma/\sqrt{n}$ is called a 95% confidence interval for the population mean. A particular interval either does or does not contain the mean, but in the long run of a large number of samples, 95% of the intervals contain the mean.

1.6.2.1 Population standard deviation unknown

The above formula for the 95% confidence interval for the population mean uses the population standard deviation, which is usually not known. In practice, therefore, we apply the formula using the sample standard deviation, s, as a substitute for the population standard deviation σ. However, to allow for estimating the standard deviation we modify the interval by replacing the value of 1.96 (the upper 2.5% point of the standardized normal distribution) with $t_{0.025,n-1}$ (denoting the upper 2.5% point of Student's t-distribution with $n-1$ degrees of freedom). Percentage points of t are obtained from statistical tables or packages, e.g. $t_{0.025,5} = 2.57$; $t_{0.025,10} = 2.23$; $t_{0.025,30} = 2.04$. The effect is to widen the interval, especially in smaller samples, to allow for the extra uncertainty in estimating the population standard deviation. Note also that the confidence intervals will vary in width from sample to sample because of random variation in the sample standard deviation s. However, the use of the multiplier t means that they are 95% confidence intervals with the property that in repeated samples 95% of the intervals contain the population mean.

Example calculation of a 95% confidence interval for a population mean
Consider the calculation of a 95% confidence interval for the mean level of
PCB in the skua population from the observed levels in a sample of size
$n = 12$ measurements. The sample statistics are as follows. Sample mean:
$\bar{x} = 20.25$ ppm; sample variance: $s^2 = 108.20$ ppm^2; sample standard deviation:
$s = 10.40$ ppm. The standard error of the sample mean is then given by
$s/\sqrt{n} = 3.00$ ppm. The upper 2.5% point of Student's t-distribution with $n - 1$
degrees of freedom is obtained from statistical tables at $t_{0.025,11} = 2.20$. The 95%
confidence interval is then $\bar{x} \pm t_{0.025,11} s/\sqrt{n} = 20.25 \pm 6.61$ ppm $= (13.64, 26.86)$
ppm

Most statistical packages have a procedure for calculating a confidence
interval for the population mean. For example, using MINITAB the command
and output are as follows:

```
MTB > tinterval 'pcb'
N    Mean   StDev   SE Mean   95.0 % C.I        Variable
12   20.25  10.40   3.00      (13.64, 26.86)    pcb
```

1.6.3 ROBUSTNESS OF METHOD FOR CALCULATING A CONFIDENCE INTERVAL

The above calculations for a 95% confidence interval apply to random samples
from a normal distribution. However, the method is based on the properties
of the sampling distribution of the sample mean, which because of the central
limit theorem approximates to a normal distribution as the sample size
increases. Thus the method can be applied to populations which are not
normal, with the proviso that in repeated samples from the same population
the calculated confidence interval will contain the population mean in *approximately* 95% of cases. The method is said to be robust to the assumption of a
normal distribution.

A more crucial assumption, however, is that the observations are selected
independently of each other. Lack of independence can lead to serious bias in
the method. In particular, if the observations in the sample are more similar
than those selected independently, then the standard error of the sample mean
will tend to be underestimated and the calculated confidence interval will be
too narrow, giving a false impression of precision.

1.6.4 CONFIDENCE INTERVALS FOR POPULATION PARAMETERS IN GENERAL

In the above sections we developed the concepts of the sampling distribution,
the standard error and a confidence interval to measure the error in using the
sample mean to estimate the population mean. The estimation of the population mean is an important problem, but the ideas apply quite generally to the

estimation of other population parameters. For example, there are methods for calculating a confidence interval for the population median and the variance. We can also calculate confidence intervals for differences between and ratios of population parameters. Details of these methods, however, depend on the particular case and are beyond the scope of this chapter. Note that not all methods are robust to the type of population distribution. For example, a confidence interval for the variance of a normal population can be calculated using the sample variance, s^2, but the method can give misleading results when applied to non-normal distributions.

A further situation in which confidence intervals are widely used is in measuring the accuracy of estimated parameters in some statistical model. Such models, which contain both systematic and random elements, play a key role in the statistical analysis of ecotoxicological data. For example, Chapter 5 discusses the linear regression model which is widely used when the relationship between two variables appears to be a straight line, apart from some random variability. The slope and intercept (i.e. the model parameters) can be estimated by fitting the line by the method of least squares, but because of the random variation the estimates will be subject to error. However, the estimation errors can be measured by using formulae for the standard errors of the estimates and calculating a 95% confidence interval of the form (parameter estimate) $\pm t_{0.025,n-2} \times$ standard error, where n is the number of pairs of data points and $t_{0.025,n-2}$ is the upper 2.5% point of Student's t-distribution with $n - 2$ degrees of freedom.

1.7 SIGNIFICANCE TESTS FOR COMPARING POPULATIONS

A significance test is the main statistical method for determining whether there is a difference between two or more populations. As noted in section 1.2.2 the problem of detecting differences arises because of the inherent variability of biological material: two samples from the same population will in general yield different measurements. This section describes the rationale of the significance test, using for illustration levels of PCB concentrations in the fat of puffins in

---

Example: levels of PCB in two puffin populations
In a study to compare PCB fat concentrations in birds of two populations of puffins the observed levels (ppm) in two random samples of birds were as follows (arranged in increasing order):

Population 1 16.00, 17.50, 23.20, 25.40, 27.80, $n = 10$
Isle of May: 32.70, 33.00, 37.40, 44.30, 46.00   Mean = 30.3 ppm

Population 2 3.00, 6.61, 7.17, 7.38, 10.30, 11.60, $n = 9$
St Kilda: 14.90, 29.40, 36.40   Mean = 14.1 ppm

two populations. In a particular application, the details of the significance test will depend on the comparison of interest and type of data, but the discussion opposite applies to a wide range of situations.

The question is: 'Do the data show that there is a difference between levels of PCB in the two populations?' The sample mean for the Isle of May is larger than for St Kilda, but is this difference larger than expected by chance because of the variation in levels between birds within each population? The significance test is a procedure for addressing this question.

1.7.1   ELEMENTS OF THE SIGNIFICANCE TEST

1.7.1.1   Null hypothesis

The null hypothesis is a specific statement about the population(s) of interest, which might or might not be true. Some examples are: (a) the population mean is equal to 10; (b) two population means are the same; (c) two populations have identical distributions; (d) the population distribution if normal; (e) two variables are independent. However, a statement that the population mean exceeds 10, is not specific enough. The null hypothesis is usually denoted by H_0, e.g. equality of populations means is written as H_0: $\mu_1 = \mu_2$.

1.7.1.2   *Test-statistic*

A test-statistic, T, is a function calculated from the sample which is used to test the null hypothesis. For example, to test the null hypothesis of equality of population means using observations in two random samples we might use the observed difference between the sample means $T = \bar{x}_1 - \bar{x}_2$. Another example, is the null hypothesis of independence between two variables x and y, for which a possible test-statistic is the sample correlation coefficient (r) calculated from pairs of (x, y) values (Chapter 2).

1.7.1.3   Statistical significance, *p*-values and significance levels

The rationale of the significance test rests on how likely would be a value of the test-statistic as or more extreme than that observed if the null hypothesis were true. The p-value is the probability of observing a value of T as or more extreme than that observed (T_{obs}) when the null hypothesis H_0 is true. The p-value measures the degree of consistency of the data with the null hypothesis. This is often broadly translated as follows: $p > 0.10$ – reasonable consistency with H_0; $0.10 < p < 0.05$ – slight evidence against H_0; $p < 0.05$ – moderate evidence against H_0; $p < 0.01$ – strong evidence against H_0. A common choice of threshold is 0.05. If the p-value ≤ 0.05, we say that the result of the significance test is 'statistically significant at 5%', or 'we reject the null hypothesis at the 5%

significance level'. A Type I error or false positive occurs when a significant result is declared when H_0 is true. Type I errors can occur by chance with error rate equal to the significance level. The above thresholds for the significance level are arbitrary and it is good practice to report the calculated p-value.

1.7.1.4 Alternative hypothesis and statistical power

In choosing a test-statistic we usually have in mind some departure from H_0 which we wish to detect, and this usually guides our choice of test-statistic. This departure from H_0 is referred to as the alternative hypothesis, sometimes denoted by H_A. Often the direction of the difference between means is not specified and the alternative hypothesis is H_A: $\mu_1 \neq \mu_2$, and we refer to this as a two-tailed test. In this case extreme values of the test-statistic are those with a large positive or a large negative value. For a one-tailed test with alternative hypothesis H_A: $\mu_1 > \mu_2$ extreme values of $T = \bar{x}_1 - \bar{x}_2$ are large and positive. The set of extreme values of the test-statistic leading to rejection of the null hypothesis is called the critical region.

The statistical power of a significance test is the probability of rejecting the null hypothesis when a specified alternative hypothesis is true, for a given significance level. For any sensible test-statistic the statistical power should increase with the magnitude of the departure from the null hypothesis. For example, in a test for a difference between two means, power should increase with the absolute difference $|\mu_1 - \mu_2|$. Power also depends on the number of observations. In practice we try to choose relatively powerful tests and design studies with sample sizes which are large enough to detect the effects of interest. A Type II error or false negative occurs when a significant result is not declared when H_0 is not true. The Type II error rate is equal to $1 -$ power.

1.7.2 STUDENT'S t-TEST FOR COMPARING THE MEANS OF TWO NORMAL POPULATIONS

The statistician W. S. Gossett (alias Student) developed the t-test for equality of means in two normal populations while working for the Guinness Brewing Company! The data are observations on a random sample drawn from each population. The test-statistic is based on the difference between the two sample means $T = \bar{x}_1 - \bar{x}_2$. The p-value is the probability of observing a value of T as or more extreme than that observed (T_{obs}), when the null hypothesis is true. A difficulty, however, is that this probability depends on the distribution of the observations. So, to make progress we further assume that: (a) the population variances are the same (i.e. $\sigma_1^2 = \sigma_2^2 = \sigma^2$); (b) the populations are normally distributed. Then, on the null hypothesis, T follows a normal distribution with mean zero and variance $\sigma^2/(1/n_1 + 1/n_2)$. The population variance is

Example: levels of PCB in two puffin populations
The calculations for Student's t-test applied to the puffin data are as follows:

Isle of May: $n_1 = 10, \overline{x}_1 = 30.33, s_1^2 = 106.09$;
St Kilda: $n_2 = 9, \overline{x}_2 = 14.08, s_2^2 = 128.14$.

The pooled estimate of the population variance is then

$$s^2 = (9 \times 106.09 + 8 \times 128.14)/(10 + 9 - 2) = 116.47$$

with estimated standard deviation $s = 10.79$.

The estimated standard deviation of the difference between the two sample means is then

$$\sqrt{[116.47/(1/10 + 1/9)]} = 4.96$$

and the observed value of the test-statistic is

$$t_{\text{obs}} = (30.33 - 14.08)/4.96 = 3.28$$

The p-value is the probability of a more extreme value, i.e. larger than 3.28 or smaller than -3.28, for a two-tailed test (section 1.7.1). From statistical tables, the upper 0.5% point of Student's t-distribution with $(n_1 + n_2 - 2) = 17$ degrees of freedom is 2.90. We conclude that there is strong evidence against the null hypothesis and that the difference in mean levels of PCB is statistically significant at the 1% level ($t_{\text{obs}} = 3.28, p < 0.01$).

Steps in the calculations for Student's t-test are given above for illustration, but in practice we can use a statistical package. The procedure and output using MINITAB are shown below.

```
MTB > twosample 'IoMay' 'StKilda';
SUBC > pooled.

Two Sample T-test and Confidence Interval

Twosample T for IofMay vs StKilda

          N     Mean    StDev   SE Mean
IofMay    10    30.3    10.3    3.3
StKilda   9     14.1    11.3    3.8

95% C.I. for mu IofMay - mu StKilda: (5.8, 26.7)
T-Test mu IofMay = mu StKilda (vs not =):
T = 3.28 P = 0.0045 DF = 17
Both use Pooled StDev = 10.8
```

An alternative procedure which allows for different variances gives very similar results:

$$T = 3.26, P = 0.0049, \text{DF} = 16.$$

Note that the output also contains a 95% confidence interval for the difference between the two population means (section 1.6.3).

usually not known so we estimate it by a weighted average of the two sample variances

$$s^2 = [(n_1 - 1) s_1^2 + (n_2 - 1) s_2^2] / (n_1 + n_2 - 2) \tag{1.15}$$

The test-statistic is then the difference of the two sample means divided by its estimated standard deviation, i.e.

$$t = \frac{\overline{x}_1 - \overline{x}_2}{\sqrt{s^2 \left(\dfrac{1}{n_1} + \dfrac{1}{n_2} \right)}} \tag{1.16}$$

In repeated random samples from two normal distributions with the same mean and variance, the statistic t follows Student's t-distribution with $(n_1 + n_2 - 2)$ degrees of freedom (section 1.5.6). The p-value (probability of observing a value as or more extreme than observed) is then obtained from statistical tables or using a statistical package.

1.7.3   ROBUSTNESS OF STUDENT'S t-TEST

A significance test is said to be robust if the Type I error is little affected by departures from the assumptions used to calculate the p-value. Student's t-test assumes normal populations with equal variances. However, the test-statistic is based on the sample means, which are approximately normally distributed whatever the distribution. So, the method is robust to departures from normality, provided that the sample sizes are not too small (at least 10). When the sample sizes are equal, the method is also robust to inequality of variances. However, there is an alternative procedure (implemented in MINITAB, section 1.7.2) based on separate estimates of variance and an adjustment to the degrees of freedom.

1.7.4   SOME ASPECTS OF APPLICATION AND INTERPRETATON OF SIGNIFICANCE TESTS

Significance tests are useful objective procedures for demonstrating differences when random variation is substantial. Some important general issues on their use are as follows:

1.   Statistical significance is not the same as biological significance. A significance test could demonstrate a difference which was small and of little consequence. On the other hand, a difference of biological significance may not be detected. This situation could arise from low power because of a relatively small sample size. Note that a result which is not statistically

significant should not be interpreted to mean that the difference is zero. A significance test can demonstrate a difference but it cannot show that there is no difference.

2. A significance test can demonstrate a difference unlikely to have occurred by chance, but it says nothing about the magnitude of a difference. In practice, therefore, we also usually estimate the difference and use a confidence interval to measure estimation error (section 1.6), e.g. the t-test and confidence interval based on the t-distribution for the difference between two means, as shown in the above MINITAB output.

3. A significance test at level 5% has probability of falsely rejecting the null hypothesis or Type I error rate of 5%, so that if several tests are applied we expect to obtain statistical significance in 5% of cases, purely by chance. This is sometimes referred to as the multiple comparison problem. An example arises in experiments to compare several treatments. Chapter 3 discusses multiple comparisons and some procedures to allow for the effect.

1.7.5 PARAMETRIC, DISTRIBUTION-FREE AND NON-PARAMETRIC TESTS

Significance tests can be classified into three broad groups reflecting the type of data and assumptions made about the distribution of the observations. In parametric tests the observations are assumed to follow a particular distribution, e.g. a normal distribution for the t-test. In distribution-free tests there are no specific assumptions about the form of the distribution. Randomization tests are examples of distribution-free tests in which the p-value is obtained from the distribution of the test-statistic under repeated random shuffling of the observations. Non-parametric tests are randomization tests applied to ranked data: the distributions of the test-statistics can be tabulated for a range of sample sizes (see Chapter 2). Distribution-free and non-parametric methods often test the null hypothesis of equality of distributions, using test-statistics chosen to have high power for detecting particular alternative hypotheses, e.g. a difference in medians. Parametric methods test hypotheses about particular parameters, and usually involve specific additional assumptions. Note that for most applications a key assumption common to all three types of test is that the observations are statistically independent of each other. In practice, the choice of test involves consideration of the null hypothesis to be tested, the nature of the data and the sample sizes.

1.8 TYPES OF STUDY: EXPERIMENTS AND SURVEYS

Aspects of the design of ecotoxicological studies are dealt with in later chapters of this book. However, from the point of view of interpretation of the results it is useful to distinguish between surveys or observational studies and experiments.

In a survey or observational study, we aim to describe the properties of a population, often using observations on selected samples of material. In an experiment we manipulate material to answer specific questions, and to establish causal links between some factor and a response. For example, the study of starlings (Appendix 3) is a survey to describe the seasonal changes in the fat, protein and metal contents found in the liver of the birds. A statistical analysis may detect differences between seasons but the study is not designed to elucidate the causes of such changes, although it could provide clues and suggest hypotheses to be tested. In the experiment on beetle behaviour, the aim is to test whether copper intake affects behaviour measured by turning rate, path length, etc. Individuals are allocated to one of four groups receiving 0 (control), 0.5, 1 or 2 units of copper, using randomization to avoid possible biases from unknown systematic differences between groups. A statistically significant difference between a treatment group and the control shows that copper causes changes in behaviour, because this is the only systematic factor which differs between treatment groups. Note that this causal inference is different to a statistically significant correlation between behaviour and levels of copper in a random sample of individuals from a field population.

Despite the differences in the aims and interpretation of surveys and experiments, the basic concepts and methods of statistical analysis discussed in this chapter apply to both types of study. For example, both approaches usually involve an element of random selection of material either through simple random sampling or randomization. Student's t-test for a difference between means applies to observations in random samples from two populations, or to the responses of individuals allocated at random to one of two treatment groups.

2

Exploring the Data

DON FRENCH[1] AND DAVID LINDLEY[2]

[1]Institute of Terrestrial Ecology, Banchory Research Station, UK
[2]Institute of Terrestrial Ecology, Merlewood Research Station, UK

2.1 INTRODUCTION

This chapter is about 'getting to know your data'. What 'shape' is a variable? How consistent is it? Are there 'odd' or extreme values, and if so, can they be explained? What effect might they have on any subsequent analyses? What do any shapes or patterns in individual variables or in their relationships with each other tell you about what your data mean? How can you best 'visualize' your data?

The better you know your way around your data, the easier it is to construct or select appropriate models for formal hypothesis testing. Indeed, the kinds of exploration described here can be very useful in generating additional hypotheses. Exploratory forays into the data also help you to avoid some of the more basic mistakes in interpretation which can occur if you go straight into formal model testing without checking if your data are suitable for the test(s) used. For example, if a test requires that variables are normally distributed, and with approximately equal variance (see Chapter 1), then if you try to apply it where these conditions are not met, the results you get may be misleading. Exploratory analyses can show up these potential stumbling blocks, and may suggest ways to avoid them, e.g. transforming the data to alter the shape of the distribution or the pattern of variation in observation scores.

In this chapter, we will discuss frequency distributions of actual data (section 2.2), how they relate to the theoretical distributions described in Chapter 1, and some of the properties of their shapes, with particular reference to applying appropriate tests in subsequent analyses and on possible transformations of the data which might be helpful (covered in more detail in section 2.3 and Chapter 3). In the next section, we describe measures of shape, measures related to frequency scores and data transformations to modify the shape or pattern of variation in the data. Section 2.4 covers ways of 'visualizing' single variables, using tables and graphs, and how well-constructed tables or graphs

Statistics in Ecotoxicology. Edited by T. Sparks. © 2000 John Wiley & Sons Ltd

can aid interpretation and analyses of the data. This is followed by an outline of ways to explore relationships between two or more variables or samples. Are there differences between groups, treatments, etc? Is there any association or correlation between two or more variables and, if so, what is the pattern of that association? Some basic statistical tests related to these concepts are described. In section 2.6, we return to 'visualization' – now looking at combinations of two or more variables, again using tables and graphs, and discussing their use in clarifying the patterns in the data. Section 2.6 deals with diversity/richness indices (a means of summarizing complex multivariate data). A final 'cautionary tale' emphasizes the importance of visualizing data, and illustrates how the techniques described in this chapter can provide a valuable guide to appropriate models and tests, and help you to avoid some of the errors caused by premature use of formal hypothesis tests.

Finally, this chapter is *not* primarily about testing explicit *a priori* hypotheses or models. Rather, it is about extracting *patterns* from the raw data, in order to

- generate or select suitable hypotheses or models to test
- ensure that the data conform to conditions required by a particular statistical test
- interpret the data, and any statistical results derived from the data, in terms of the actual mechanisms or relationships operating in the systems under study.

2.2 FREQUENCY DATA AND FREQUENCY DISTRIBUTIONS

2.2.1 GENERAL CONCEPTS

Any variable which is measured on a quantitative scale – i.e. where the observation scores represent numerical values – can be expressed as a frequency distribution. The scores used need not be on a continuous scale. Indeed if they are on a continuous scale, they must be grouped in some way to enable a frequency distribution to be compiled. For example, an experiment to test the effects of a potentially toxic substance on vegetation might use the heights of plants of known age as a measure of growth rates (a continuous scale, needing to be grouped into height classes), counts of seeds germinating in pots (integer values, no automatic need to group) or estimates of vegetation cover measured to the nearest 5% (a discontinuous scale, but only because of imprecise measurement – in principle, representing a continuous underlying scale). For exploratory purposes, scores need not even be on an interval scale, but there must be some 'natural' ordering in the scoring. Vegetation classes (grassland, heath, bog . . .), for example, are nominal variables and have no natural ordering in themselves – a value of 'grassland' for vegetation class is in no sense 'higher' or 'lower' than one of 'bog' – but the sequence 'open moor', 'scattered trees', 'open-canopy forest', 'closed-canopy forest' represents an ordered set of

Table 2.1 Frequency distributions of body lean weight (g) and cadmium concentration (mg kg^{-1} DW) in 57 starlings. Variables are truncated to the nearest lower integer for allocation to class intervals in the table

Body lean weight	20	21	22	23	24	25	26	27	28	29		
Frequency	1	6	7	14	11	5	5	6	1	1		
Cd	0	1	2	3	4	5	6	7	8	9	10	11
Frequency	1	5	17	11	10	5	3	2	1	1	0	1

scores for tree density, so could alternatively be expressed as numerical rank scores for tree density, for subsequent calculation of statistical summaries and comparisons.

The data on 57 starlings (Appendix 3) provides a convenient illustration. Body lean weights (g) range from 20.68 to 29.31 so division into integer intervals (20–21, 21–22, etc.) gives 10 ordered weight classes. Cadmium concentrations (mg kg^{-1} dry weight (DW), in the same set of birds, range from 0.79 to 11.38 so grouping into integer intervals (0–1, 1–2, etc.) gives 12 classes. The two frequency distributions are listed in Table 2.1.

Some features of the two distributions are immediately apparent even from this simple tabulation. Both distributions have frequency rising from low values to a peak, then steadily falling off again at higher values. So both distributions imply the existence of some kind of modal value, with observations tending to cluster around it. However, the peak frequency in body lean weights is around the middle of the distribution, at 23–24, and the distribution is approximately symmetrical about these central values, with only a slight tendency to a 'tail' at the upper end. In contrast, cadmium concentrations rise rapidly to a peak frequency at 2, with a long gradual tailing-off in a large number of higher classes. These features of symmetry and asymmetry (or skewness), and the degree to which values cluster close to some average value (kurtosis), are discussed in more detail in the next section.

2.2.2 SYMMETRY AND SKEWNESS, KURTOSIS AND SPREAD

We have just observed that the body lean weight is more or less symmetrical (i.e. not skewed), whereas cadmium levels are positively skewed (a long 'tail' to the distribution in the higher classes). Negative skew would show the opposite pattern, with a peak near the top end of the range of values, and a long tail in the lower classes. Figure 2.1 gives a generalized illustration of these three basic shapes.

A related concept is that of kurtosis, or the degree to which values are clustered closely about some central or average value. A distribution with high kurtosis has values clustered tightly about an average value; low kurtosis, conversely, implies fewer values close to the average, with a corresponding

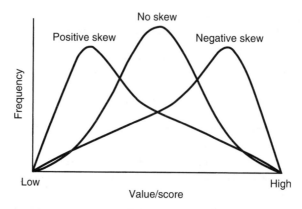

Figure 2.1 Shapes of frequency distributions with different types of skew.

wider 'spread' of values. Different degrees of kurtosis are illustrated in Figure 2.2.

Skewness and kurtostis are both important considerations when making inferences about, for example, effects of experimental treatments, or differences between classes or groupings of survey data. There are two ideas to consider here. Firstly, if differences are found between treatments, their interpretation is likely to depend on whether the differences are expressed in average values (however defined), or are mainly differences in the extent of tails in highly skewed distributions. In the former case, differences between groups are likely to apply to a group as a whole. In the latter case, it may be that only a subset of 'susceptible individuals' are involved. Similar arguments apply to considerations of spread. There may well be cases where a response variable shows no change

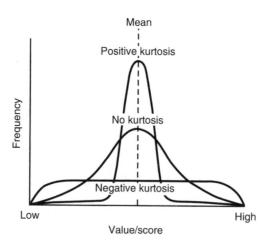

Figure 2.2 Shapes of frequency distributions with different degrees of kurtosis.

in its average value as a result of a treatment, but does increase or decrease in variability. The second consideration is that several statistical tests assume certain characteristics of the underlying distributions of the variables measured (for example, that they conform approximately to a normal distribution, or that they are symmetrical), and if these assumptions are not met, the test may not be valid for those data. If, however, the deviation from the assumed form of the distribution is known, it may be possible to transform the data into something closer to the assumed form, for example by taking square roots or logarithms. Transformations are discussed further in section 2.3, and treated more extensively in Chapter 3. Here we merely note their potential use in modifying the shape of a frequency distribution.

2.3 SUMMARY MEASURES AND TRANSFORMATIONS

The preliminary examination of frequency distributions described in the previous section is a start in reducing the initial chaos of a raw, unordered, dataset to something more coherent and manageable. A second approach to getting a 'feel' for the data is by the use of summary measures. These provide estimates of some quantitative properties of the sample or distribution, such as its 'average' value, or the most common score, the amount of variation about the average point, what is the pattern of that variation, etc. Examination of combinations of these summary measures can tell us a lot about the overall shape of the data, and may indicate whether particular transformations of the data are likely to be useful.

Summary measures can be divided into three groups: measures of average value, measures of variation about an average, and measures of shape related to frequency distributions. We will first discuss each of these, with examples of the more commonly used measures, then outline the situations where transformations may be useful, how these may be recognized using frequency distributions and summary measures and briefly describe a few commonly used transformations.

2.3.1 MEASURES OF 'AVERAGE' VALUE

The mean, mode and median have already been introduced in Chapter 1. An alternative mean measure is sometimes used – the *geometric mean*, defined as the nth root of the values multiplied together, i.e.

$$\sqrt[n]{(x_1 \times x_2 \times \ldots \times x_n)} \qquad (2.1)$$

The geometric mean is used for the calculation of means of sequences whose individuals occur at a definite constant ratio. However, it is worth noting here that if a variable is log transformed, then calculating the arithmetic mean of the log-transformed data is equivalent to calculating the geometric mean of the untransformed values. It is important to bear this in mind if you apply a logarithmic transformation to any data.

For the starling body weight data the arithmetic and geometric means are broadly similar at 24.34 and 24.26 g respectively.

2.3.2 MEASURES OF VARIATION

Chapter 1 has already introduced the concepts of variance, standard deviation, standard error of the mean and coefficient of variation.

Table 2.2 summaries the means and variability of four variables contained in the starling dataset. The body lean weight shows little variability, while more variability is present in the manganese and cadmium concentrations found in the starlings. However, this table highlights the fact that the distribution of mercury follows a more extreme pattern than any of the previous variables. Its coefficient of variation is very high. Examining the original data shows why; the mean is heavily weighted by 31 low values of 0.08 (the detection limit), but the variance is influenced by the 'tail' of this highly skewed distribution, including some very high values (6.39, 7.05, 8.62, 12.1).

It is important to repeat that the standard error of the mean measures a different kind of variation to both variance and standard deviation. Variance and standard deviation are both measures of the variation in the *population* from which the sample was taken. They do *not* vary intrinsically with the size of the sample, so samples of different sizes from the same population should give broadly similar results when variance or standard deviation are calculated. The standard error of the mean, however, is a measure of how precisely the mean is measured *in a particular sample*. Thus, the larger the sample from a given population, the smaller the standard error of the mean. When comparing standard errors of means, therefore, it is essential to know what the sample sizes are, and these should always be quoted with standard errors.

The ideas of asymmetry or *skewness* (a long tail in one direction of the

Table 2.2 Measures of the mean and variability of four variables from the starling dataset ($n = 57$)

	Mean	Variance	Standard deviation	Standard error of the mean	Coefficient of variation (%)
Body lean weight (g)	24.34	3.86	1.96	0.26	8.1
Mn (mg kg^{-1} DW)	5.01	5.63	2.37	0.31	47.3
Cd (mg kg^{-1} DW)	3.94	4.10	2.02	0.27	51.4
Hg (mg kg^{-1} DW)	1.00	5.06	2.25	0.30	225.1

distribution) and *kurtosis* (the degree to which data are concentrated around the mean value) have been described qualitatively in relation to frequency distributions (section 2.2.2, above). Both properties of a distribution can also be summarized quantitatively, as extensions of the formula for variance. Skewness is measured as the average of cubed deviations from the mean. Kurtosis is similarly derived from the fourth-power deviations. Mathematically:

$$\text{skewness} = \Sigma\,(x_i - \bar{x})^3/(n - 1) \tag{2.2}$$

$$\text{kurtosis} = \Sigma\,(x_i - \bar{x})^4/(n - 1) \tag{2.3}$$

A positive value for skewness indicates that the tail is towards the positive end of the distribution, a negative value that the tail is towards the negative end. A zero value indicates perfect symmetry. Similarly, a high positive value for kurtosis reflects a high proportion of values 'clumped' near the mean, a negative value a wide spread of observations, and a zero value the degree of clumping about the mean expected from the bell-shaped normal distribution (see Chapter 1).

Table 2.3 gives the values for skewness and kurtosis for five variables from the starling dataset. There is increasing skewness to the right in the variables as you move down the table, and also increasing kurtosis. The near-zero values for both skewness and kurtosis in body lean weight shows that the distribution of this variable is close to symmetrical and with a degree of aggregation about the mean similar to a normal distribution. In contrast, the high values of skewness and kurtosis for mercury show clearly that the measurements for this variable are heavily concentrated around the mean, but with a long 'tail' to a few birds with extremely high concentrations of mercury in their bodies.

2.3.3 'SHAPE' MEASURES RELATED TO FREQUENCY SCORES

As well as summary measures for the average value of a variable, and the size and shape of the variation about it, derived from the values of individual observations, there are several useful measures which summarize the pattern of

Table 2.3 Measures of the 'centre' of a distribution in relation to skewness and kurtosis (variables from the starling dataset)

	Mean	Median	Mode	Skewness	Kurtosis
Lipid concentration in liver (mg/g)	64.00	56.00	28.57	0.25	−1.03
Body lean weight (g)	23.34	24.11	24.13	0.43	−0.40
Cd (mg kg^{-1} DW)	5.01	4.62	4.65	1.40	2.61
Mn (mg kg^{-1} DW)	3.94	3.45	1.95	1.75	4.45
Hg (mg kg^{-1} DW)	1.00	0.08	0.08	3.35	12.97

scores, in a way directly related to the overall frequency distribution. In some cases, these may be more informative, or better to use for statistical tests, than measures such as mean and variance, particularly if the data are counts of observations with discrete or discontinuous values.

Mean, median and mode can all be used as measures of the 'centre' of such distributions, in the same way as continuous distributions, but while the mode is generally of little use in continuous distributions measured accurately (Table 2.3), if data are discontinuous, or can be grouped into discrete class-intervals as in Table 2.1, it can be very useful to identify not only the true modal class, but also any subsidiary peaks in the distribution, as these may identify distinct subsets of the population. You might like to try grouping the values for lipid in liver from the starling dataset into 20 mg class intervals, and identify the modal classes (see also section 2.4.2).

There is no direct equivalent of variance and its derivatives, or of skewness and kurtosis, in frequency measures. However, a similar overall picture of the shape of the data can be built up using the range and percentiles of the distribution.

The range is simply the difference between the minimum and maximum values of a variable. Alternatively, the actual minimum and maximum can be presented.

Percentiles are those values defining the upper limit of the values of some given percentage of the observations. For example the 10th percentile is that value where 10% of the observations are smaller and 90% are greater. Conversely, the 90th percentile has 90% of observations smaller and 10% greater. These two percentiles, together with the quartiles (25th and 75th percentiles) and the median (the 50th percentile) are commonly used to build up an idea of the overall shape of a distribution, together with the minimum and maximum values. (You can, however, use any thresholds you like if others are more applicable to your particular datasets.) Similarly, inter-percentile ranges (e.g. inter-quartile range, or from the 10th to the 90th percentile) together with the overall range, can be used to describe the degree of variability in the data about the median, in a way analogous to the use of the standard deviation to describe the variation about a mean value. Table 2.4 gives some examples from the starling dataset, which you may wish to compare with the measures in Tables 2.2 and 2.3. Generally, the ranges (inter-quartile, etc.) are not by themselves as informative as variance, standard deviation, etc., but together with the actual values for the percentiles, provide a good overall picture of the pattern of variation in the data.

2.3.4 TRANSFORMATIONS

Many statistical tests assume that there are particular patterns of variation in the data, for example that the populations from which samples are taken have a

Table 2.4 Percentiles and ranges as measures of shape and variability in a frequency distribution (data from the starling dataset). Hg data minimum, up to median, are 0.08 rounded up

	Percentiles						
	Minimum	10%	25%	Median	75%	90%	Maximum
Liver weight	2.3	2.7	3.5	3.9	4.2	4.6	5.3
Body lean weight	20.7	21.8	23.2	24.1	25.8	27.3	29.3
Cd	0.8	2.0	2.5	3.5	4.8	6.6	11.4
Mn	0.5	2.8	3.5	4.6	5.7	8.6	14.3
Hg	0.1	0.1	0.1	0.1	0.7	2.4	12.1

	Ranges			
	Median	Inter-quartile	10%–90%	Min–max
Liver weight	3.9	0.8	1.9	3.1
Body lean weight	24.1	2.5	5.6	8.6
Cd	3.5	2.3	4.7	10.6
Mn	4.6	2.2	4.9	13.9
Hg	0.1	0.7	2.3	12.0

normal distribution (see Chapter 1) or that the variances of two samples are equal or that a sample distribution is symmetrical about its mean. These assumptions may not be satisfied in a particular sample, in which case you have, broadly, two alternative approaches. Firstly, you can use only *non-parametric* tests. These make no assumptions about the form of a distribution – some are described later in this chapter (see section 2.5), others are dealt with in Chapter 3. Siegel and Castellan (1988) give detailed descriptions of all the more commonly used non-parametric tests. However, while these tests can be used with data of virtually any distribution, they are often less powerful (i.e. less capable of detecting real differences or associations) than the equivalent *parametric* tests (i.e. tests which make one or more assumptions about the form of the sample distribution). Also, there are some kinds of analyses for which non-parametric tests do not exist. That leaves the second possibility – transforming the data so that the distribution or pattern of variation conforms to the assumptions of the test. Transformations are discussed in Chapter 3; here we simply draw attention to some of the more common transformations and their application.

There are two main reasons for transforming data: firstly, to convert the shape of a distribution to one conforming more closely with the requirements of a particular test, and secondly, to equalize the within-sample variation over several samples when they are being compared with each other.

Common transformations are square root (used when the variances of samples vary proportionally with their means), logarithmic (when standard deviations of samples vary proportionally with their means, or to reduce positive skewness in a distribution and so render it more symmetrical), and angular transformations (used to normalize the pattern of variance when data are proportions or percentages).

There are two general problems with all these transformations. Firstly, some data values may not be directly transformable; none of the above transformations are possible on negative values, there is no 'real' value for log(0), and angular transforms are not possible for values over 100%. This problem can in many cases be circumvented by simply adding a constant to all values, or otherwise rescaling them to bring all values into a valid range. However, this may complicate interpretation, since the data now have a shifted baseline. Secondly, means of transformed variables are usually not the same as those from the untransformed data, and this must always be allowed for when interpreting the results of any analysis in terms of the original measurements (e.g. setting 'acceptable' threshold values for a pollutant).

The square-root transformation is probably the simplest, both conceptually and in its effects. It is calculated by taking the square root of each observation value.

Logarithmic transformations are at first sight as simple in principle, but more complex in practice. As well as the two general problems of untransformable values and changing patterns of means, logarithmic transforms also alter the meaning of basic arithmetic operations, when applied to the original data. We have already referred to the arithmetic mean of log-transformed data being equivalent to the geometric mean of the untransformed data. The same principle applies more generally; adding transformed values is equivalent to multiplying the untransformed values, subtraction to division, multiplication to exponentiation, etc. and can easily result in incorrect model or test selection, misleading results or erroneous interpretation, if particular care is not taken over this point.

Of the many possible angular transformations, the most commonly used is arc sine (square root). This is calculated by first converting the data to proportions (if they were originally percentages), then taking square roots, then calculating the arc sines of the resulting values. This transformation is simple to calculate, but can give difficulties in interpretation, not least in the conceptual shift from a percentage score to an angle in degrees or radians! There is also some confusion in published accounts between arc sine and arc sine(square root) – in most statistical tables it is the latter that is assumed, but it is frequently referred to simply as 'arc sine' (omitting the 'square root') in the same way as 'mean' is usually taken to be 'arithmetic mean'. In practice, angular transformations have little effect if the data lie mainly in the range 30%–70%.

No matter what transformation may seem appropriate for your data, there is one final general caveat. It may be that there is no standard (or indeed non-standard) transformation which will convert your data into the form required by a given test. This is particularly so when the sample distribution is extremely heavily skewed or bi- or multimodal or otherwise 'odd'. It may also be the case that no transformation will adequately homogenize the variances of all the samples you may wish to compare. If you run into either of these difficulties, you should consider whether:

1. There is a suitable non-parametric test which will answer the question you wish to ask of the data. Or
2. There is any parametric test which does *not* make the particular assumption that your data may be violating. Or
3. The data allow valid statistical inferences to be drawn at all on the questions you want to ask. It is not impossible for a dataset to allow no valid statistical analyses relevant to particular questions. If you are unsure about this, we can only advise you to consult a statistician!

2.4 VISUALIZING THE DATA I: SINGLE VARIABLES

Visualization, or presenting data in a way whereby they can be scanned by eye, and patterns and shapes easily seen, is an extremely important facet of data exploration. Indeed, we consider it to be so important that we deal with it three times in this chapter: first, in this section, ways of assessing the internal shapes or patterns in individual variables; secondly, in section 2.6, how to use tables and graphs to examine relationships between variables, interpret differences between groups or samples and see patterns of association or correlation; thirdly, in the final section, we give an illustration of the general importance of visualization in the exploration of data, and how it can help you to avoid making unnecessary errors by prematurely launching into formal hypothesis testing before thoroughly exploring the underlying patterns. Tables are generally more useful than figures when the actual quantities are of interest; graphs tend to be better when the shapes or patterns are of prime importance.

2.4.1 TABLES AND LISTS

Raw data are usually pretty messy in their original state. Seeing any clear patterns in most of the variables in the starling dataset is quite difficult. However, if the data are presented as an ordered list, some features immediately become clearer. For example, if the cadmium concentrations are listed in order from lowest to highest, we obtain:

0.79	1.35	1.47	1.89	1.95	1.95	2.02	2.08	2.13	2.27	2.30	2.41
2.45	2.46	2.53	2.56	2.56	2.73	2.78	2.82	2.91	2.93	2.93	3.05
3.06	3.30	3.39	3.45	3.45	3.49	3.56	3.59	3.73	3.76	4.09	4.17
4.28	4.37	4.40	4.55	4.62	4.81	4.82	4.93	5.12	5.16	5.31	5.58
5.92	6.06	6.10	6.62	7.29	7.39	8.12	9.23	11.38			

Already it is easier to see a pattern; there are few values below 2.0, quite a lot at 2–3, slightly fewer at 3–4, then fewer and fewer at successively higher values, up to the maximum of 11.38. Rounding the data, say to nearest lower integer, so as not to obscure general trends by excessive detail, might give the following:

0	1	1	1	1	1	1	1	2	2	2	2	2
2	2	2	2	2	2	2	2	2	2	2	2	3
3	3	3	3	3	3	3	3	3	3	4	4	4
4	4	4	4	4	4	4	5	5	5	5	5	6
6	6	7	7	8	9	11						

The pattern becomes clearer still.

Grouping the list in these integer classes,

```
0
1  1  1  1  1
2  2  2  2  2  2  2  2  2  2  2  2  2  2  2  2  2
3  3  3  3  3  3  3  3  3  3  3
4  4  4  4  4  4  4  4  4  4
5  5  5  5  5
6  6  6
7  7
8
9
11
```

we are approaching a graphical representation (see also section 2.4.2) and the pattern is clearer still. The above presentation is similar to a bar chart or histogram (see next section). An alternative way of expressing this last listing is in the form of a frequency table, as was done for this variable in Table 2.1 – again using integer classes to group the data.

There are three basic principles involved in the progressive clarification of the shape of a dataset by means of tables and lists, as exemplified above:

1. Ordering the data in some logical sequence – usually from lowest to highest values or vice versa.
2. Presenting the values at an optimal precision. As a general rule, consider carefully if it is really necessary to give data in a summary table such as those above, to more than two significant figures at most – usually it is not, and the lower precision of individual values clarifies the general patterns.
3. Grouping the data values into appropriate class intervals to create a frequency table. Here it is important to select an appropriate class interval – too few, or too many, within the range of the data, and the pattern is lost. Some examples illustrating the effects of different class sizes on the patterns shown are given in the next section.

2.4.2   HISTOGRAMS, BAR CHARTS AND FREQUENCY CURVES

It is often useful to make a graph of a frequency distribution, in the form of a histogram or bar chart, or as a frequency curve. The last example in the previous section is effectively a kind of bar chart of the cadmium values, since the length of each row is proportional to the number of observations in that class interval. This is not quite true in the last row, where a bar chart should have a blank row at 10, and a single character width at 11. A histogram is similar to a bar chart, but strictly, in a histogram, it is the area of each column or bar which is proportional to the frequency or score, rather than the height or length. This makes a histogram, *sensu stricto*, more appropriate when the class intervals are not all of equal size. The two terms are in practice quite frequently used loosely as interchangeable alternatives, but we recommend that you keep the distinction clear. Bar charts can also summarize measures other than frequency.

Figure 2.3 gives some examples of histograms of some of the variables from the starling dataset. It is immediately clear from these that moult score is highly negatively skewed (most birds had moult score 34), and mercury highly positively skewed, while body lean weight, in contrast, is approximately symmetrical. The

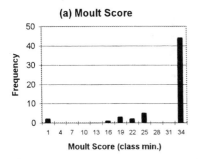
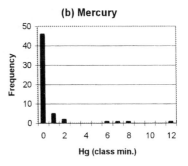
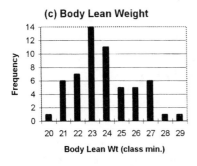
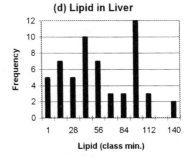

Figure 2.3   Examples of histograms/bar charts showing different kinds of sample frequency distributions. (a) Moult score (strong negative skew); (b) mercury (strong positive skew); (c) body lean weight (approximately symmetrical distribution; (d) lipid in liver (bi- and multi-modal).

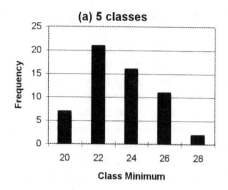

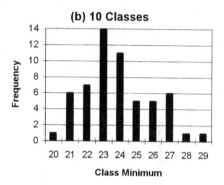

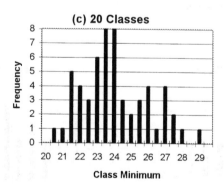

Figure 2.4 Histograms/bar charts of body lean weight showing the effect of using different class sizes on the apparent shape of the distribution. (a) Five classes each size 2 (20–22, 22–24, etc.); (b) 10 classes each size 1 (20–21, 21–22, etc.); 20 classes each size 0.5; (c) (20 20.5, 20.5–21, etc.).

histogram for lipid in liver shows an interesting pattern, with an apparent asymmetry which might not be expected from its near-zero skewness score (Table 2.3), as well as showing some of the causes of the negative kurtosis score. The apparent bimodality of the distribution, with a broad peak in the lower part of the range (around 20–60) but also a high sharp peak around about 100, in the middle of what would otherwise be a positively skewed 'tail', would not be obvious from any set of summary measures, unless the frequency distribution had been scanned for all modal classes, and it is probably worth noting here that computer packages rarely report more than a single mode, so they would *not* pick out this point. Only a visual examination of the data, either graphically as here, or using a frequency table, would reveal this aspect.

As with tables, it is important when drawing a histogram or bar chart to choose suitable class-interval sizes. The effects of three different class sizes on the apparent shape of the variable 'body lean weight' are illustrated in Figure 2.4. With only five classes (20–22, 22–24, etc.) spanning the range, the distribution appears slightly skewed with a single peak at about 23. With integer classes (20–21, 21–22, etc.) it is less definitely unimodal – as well as the main peak at 23–24 there may be a small secondary peak at about 27–28. At class intervals of 0.5, while there is still a clear peak at about 24 (note, incidentally, the gradual 'slippage' upwards of the modal class value), the 'normal'-looking bell shape seems to be breaking up, and there are at least two secondary peaks at 21.5, and at about 26.5. Generally, the overall shape of any sample distribution tends to fragment with too many classes, while with too few, any important details of its shape will be blurred, and only really gross features will be visible.

An alternative way to express the data is as a frequency curve (linking class midpoints with either a line or a smoothed curve). The shape of a distribution may be expressed more clearly thus than in a histogram or bar chart. Figure 2.5 shows some examples linking class midpoints.

2.4.3 BOX AND WHISKER PLOTS

Box and whisker plots (or 'boxplots' for short), are the graphical equivalent of the summary measures described in section 2.3. One common form of these is for a box to represent the interquartile range with 'whiskers', extending beyond the box, to represent outlying points. Within the box the position of the median is indicated. Variants on this arrangement include the replacement of the median with the mean, and hierarchical boxes to indicate different features such as interquartile range and 10th–90th percentile ranges, or standard error and standard deviation.

Figure 2.6 illustrates one form of boxplot for some variables from the starling dataset. Like histograms, they can give a good visual impression of the overall patterns of variation in the data. In this versions outlying values are represented by asterisks. The position of the median within the interquartile range box, and the relative lengths of the whiskers indicate the symmetry of the distribution.

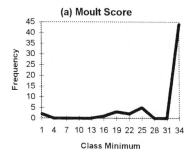

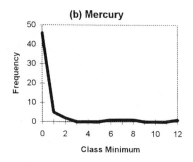

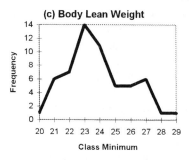

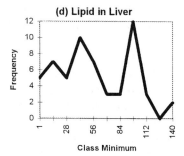

Figure 2.5 Frequency curves for the variables used in the histograms of Figure 2.3 – an alternative method for showing different patterns of frequency distributions. (a) Moult score; (b) mercury; (c) body lean weight; (d) lipid in liver.

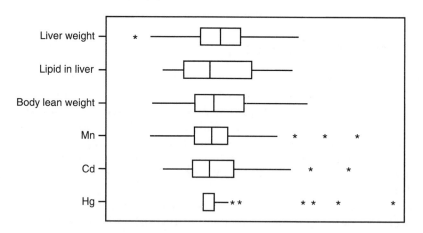

Figure 2.6 Boxplots to show examples of different kinds of variables from the starling dataset. All variables scaled to same mean and variance. Outliers for this version are defined as more than 1.5 times the interquartile range beyond the quartiles.

Liver weight, for example, is broadly symmetrical (median central within box, whiskers equal length) with the exception of a single outlier. In contrast mercury is very skew (as noted earlier), with minimum, lower quartile and median all equal and a number of higher outliers.

The various tables and diagrams described can all be useful in getting a general appreciation of the form of your data, by enabling you to scan a dataset visually, and see directly such features as skewness, amount of variation, and where the mean (or other 'central' point) lies in relation to the overall range. In short, they enable you to make an efficient assessment of the 'shape' of a variable. Additionally, they can give you some clear pointers to potential trans-formations, for example by showing where the shape of your data differs from whatever theoretical distribution is assumed by your proposed test(s). And, of course, you can easily check the effects of any transformation by simply plotting the transformed data.

2.5 RELATIONSHIPS BETWEEN VARIABLES AND SAMPLES

So far, we have considered only a single variable at a time, measured on a single sample. We now need to explore the situation where we have several variables measured on a sample (e.g. mercury, cadmium, etc.) or the same variable measured on two or more samples or subgroups (e.g. male or female starlings). Are different variables related to each other? Do different samples have different levels of a variable? Are there other insights we can gain by grouping sets of variables or observations together?

One way to think of what follows is as another kind of summary measure. This time, it is a summary of some chosen feature(s) of a combination of samples or variables, rather than of single samples or single variables. We have thus moved back another step from the original data, to get a broader view of the underlying patterns. As before, discussion of the relevant statistics will be followed by a section on visualising these combinations of variables or samples.

2.5.1 RELATIONSHIPS IN ONE VARIABLE BETWEEN MEASUREMENTS FROM TWO OR MORE SAMPLES

The dataset of 57 starlings are the results from a single sample taken from one area of countryside. Another sample from another area would produce a different dataset. These two datasets could be summarized as

Location	Mean body weight	SE	n
Area 1	24.34	0.26	57
Area 2	26.56	0.39	43

Looking at this summary raises the question 'have these two samples of starlings been collected from the same population or have they been gathered from two different populations?' i.e. does the difference between the mean values reflect a 'real' difference between starlings from the two areas? This question can be answered by testing the null hypothesis that there is no difference between the means of the two samples.

An important point to understand at this stage is that the selection of an appropriate statistical test is of paramount importance. With ready access to a wide range of statistical packages it is only too easy to enter your dataset into a package and get out an 'answer' – but if you have applied an inappropriate test, that answer is likely to be wrong!

We noted earlier that one of the purposes of extracting patterns from the raw data was to ensure that the data conform to any conditions required by a particular statistical test. We also saw earlier that the normal distribution is characterized by its parameters of a mean and standard deviation. There are a large number of statistical tests that make assumptions about the distribution of the data in terms of the parameters describing the distributions. These are called parametric statistical tests. In contrast, there is another group whose tests require no direct assumption about the value of any parameter in the underlying probability distribution. This second set of tests are called non-parametric or distribution-free tests. Table 2.5 summaries a few of the tests that can be used in three different situations. Chapter 1 introduced t-tests and analysis of variance will be covered in Chapter 3. Non-parametric tests can be useful when first exploring your data so these tests will now be considered with an example.

We have space here for only very brief introductory descriptions. For more detailed treatment of these and other related tests, see *Nonparametric Statistics for the Behavioural Sciences* (Siegel and Castellan 1988), which also contains an extensive set of tables of critical values needed for hypothesis testing.

2.5.1.1   Sign test for paired samples

To illustrate this test, we have concocted an example. Suppose we had measured some contaminant in the same birds at two time points, for example cadmium

Table 2.5   Tests of differences between the centres of sets of data

Data structure	Parametric	Non-parametric
Two groups (paired samples)	t-test	Sign test
Two groups (independent)	t-test	Mann–Whitney
>Two groups	Analysis of variance	Kruskal–Wallis One-way analysis of ranks

Table 2.6 Differences in cadmium levels in birds between two sample times

Cd at start	Cd after six months	Difference	Sign
1	4	3	+
2	3	1	+
2	7	5	+
4	4	0	0
2	6	4	+
3	6	3	+
2	4	2	+
2	3	1	+
3	5	2	+
3	6	3	+
3	4	1	+
2	8	4	+
2	2	0	0
3	4	1	+
5	4	−1	−
5	3	−2	−
11	3	−8	−
3	4	1	+
3	4	1	+
5	3	−2	−
5	2	−3	−

from blood samples taken six months apart. Then instead of having two unrelated samples we have paired samples. The pairing or matching of samples in this way is an essential condition for the sign test or any other paired-sample or matched-sample test.

The sign test examines whether the *direction* of difference is sufficiently consistent to indicate a real change over time. Table 2.6 lists the differences between the two sample times, and shows whether this difference is positive or negative. In Table 2.6, we have 14 + and 5 − signs, and two pairs with no difference. If an individual shows no change, it does not contribute to the test, so is excluded from the analysis. That leaves 19 pairs showing change. For the sign test the test statistic (Y) is the number of whichever sign occurs less often, in this case the number of negative signs. The binomial distribution is used to examine how extreme the above result is given a null hypothesis of equal numbers of plus and minus signs, i.e. Ho:$p = 0.5$.

Table in Siegel and Castellan (1988) gives a set of probabilities associated with values as small as observed values of the test statistic; N in table is the total number of observations with any non-zero sign. For $N = 19$, $Y = 5$ has a one-sided probability of occurrence under the null hypothesis of $p = 0.032$. The corresponding two-tailed probability is $p = 0.064$. These results indicate a tendency for cadmium levels to increase between the two sample times.

2.5.1.2 Wilcoxon–Mann–Whitney rank sum test for two independent groups

This is a test for difference between two *independent* samples, i.e. where the two groups of observations are not matched or paired in any way. It is often used when the data may not satisfy the assumptions of the *t*-test or when measurements are not on an interval scale (see Chapter 1 for description of different kinds of measurement scales). It was developed independently by Wilcoxon, and by Mann and Whitney, so may be referred to in other texts, or in computer packages, as the Wilcoxon rank sum test or as the Mann–Whitney *U*-test.

To explain the test, we return to the cadmium measures from the starling data set. We will test whether birds sampled in months 4 + 5 contain more cadmium than birds sampled in months 1 + 2. Note that this hypothesis does not assume anything about changes in individual birds, unlike the paired samples used in the sign test example. Instead, it is an equivalent hypothesis about changes in the average level of cadmium in the bird population.

To calculate the test, we first assign ranks from 1 to *n* (the number of observations). Where there are two or more equal measurements, they are given the average of the ranks they cover. Table 2.7 gives a summary of rank scores, ordered by month of record. From this table it can already be seen that most of the highest ranks are from the later sample times, and the lowest ranks from the earlier times. To test whether the difference in cadmium levels in starlings between the two sample times is statistically significant, we simply calculate the

Table 2.7 Rank scores for cadmium levels in starlings in different months

Month	Cd	rank
1	1.35	1
1	1.95	2.5
1	2.08	3
2	2.27	4
2	2.45	6
2	2.78	9
2	4.4	13
4	1.95	2.5
4	2.3	5
4	2.46	7
4	2.53	8
4	3.39	12
4	3.3	11
5	2.91	10
5	4.93	14
5	5.16	15
5	11.38	16

sum of ranks, Wx, for the smaller group. This gives 38.5 for months $1 + 2$. The numbers of observations in each group, m and n, are 7 and 10 respectively. For small samples such as this, we can look up Wx directly in appropriate tables (Siegel and Castellan 1988) where we find the one-tailed probability of a value as small as 38.5 is less than 0.0093, so the difference is highly significant.

For larger samples, where one or both groups contain more than 10 observations, the distribution of Wx approximates closely to the normal distribution, and a conversion to normal probability tables is possible. See Siegel and Castellan (1988) for details.

2.5.1.3 Kruskal–Wallis one-way analysis of variance by ranks

This test helps you check whether several independent samples originate from similar populations. For example, Table 2.8 shows the mean iron concentration in the livers of starlings sampled in different months (means rounded to nearest integer). There is an obvious difference between the first 6 months and the later ones, and there might also be a difference between months 7–10 and months 11–13. We might know of changes in the environment of the starlings which would correspond with this division into three groups, and want to test if the groups were statistically different. The procedure for the Kruskal–Wallis analysis is as follows.

First, rank all the observations together, from 1 (smallest) to N (largest), assigning average ranks to tied observations as usual and calculate the mean rank, R. Now arrange the ranks into k groups (in this case $k = 3$, the three time-periods) and calculate the mean rank, R_i, and the number of observations, n_i, for each group. The relevant values in this example are: R (overall mean rank) = 28.82, R_1 (months 1–6) = 17.05, R_2 (months 7–10) = 38.71, R_3 (months 11–13) = 31.47, $n_1 = 21$, $n_2 = 21$, $n_3 = 15$. The test statistic, KW, is then calculated as

$$KW = \frac{12}{n(n + 1)} \times \Sigma n_i (R_i - R)^2 \qquad (2.4)$$

In this example, that is

$$\begin{aligned}(12/(57 \times 58)) \times ((21 \times (17.0 - 28.8))^2 + (21 \times (38.7 - 28.8))^2 \\ + (15 \times (31.5 - 28.8))^2) = 0.00363 \times (2912 + 2054 + 105) = 18.4\end{aligned} \qquad (2.5)$$

If there are three or fewer groups and all have five or fewer observations, an exact probability can be found in tables of Siegel and Castellan (1988). For larger samples (as here), KW is approximately equal to χ^2 (chi-square), and can be compared with tables of critical values of χ^2 with $k - 1$ degrees of freedom

Table 2.8 Mean iron content of starlings in different months

Month	1	2	4	5	6	7	8	9	10	11	12	13
Fe(mg kg^1 DW)	17	24	19	17	12	32	27	42	36	26	33	24

(d.f.). In our example, $k = 3$, so we look up χ^2 with 2 d.f. The value of 18.4 has an associated probability of < 0.001, we can therefore confidently reject the null hypothesis that all three groups of starlings were from similar populations, i.e. they differ in their Fe concentrations. However, this test only tells us that there is a difference somewhere among the three groups. Additional tests would be needed to identify where any differences were, for example is the difference between the first period and the other two the only 'real' difference, or are all three groups different from each other? Siegel and Castellan give details of various extensions of the Kruskal–Wallis test, including locating where the differences are.

2.5.2   ASSOCIATIONS AND CORRELATIONS

For the purposes of this chapter, we will refer to relationships between categorical variables (e.g. site type, animal presence/absence) as 'associations' and those between interval variables (e.g. levels of a pollutant, animal numbers) as 'correlations'. Although both are measures of agreement between two variables, they are used to test relationships between different kinds of measurements.

There are a large number of tests of association and correlation. We have space here only for a very brief introduction to the more common ones. For details of other tests, and extensions of those described here, we refer the reader to Siegel and Castellan (1988), or Sokal and Rohlf (1995).

2.5.5.1   Associations

Suppose we are concerned about a specific beetle that can be found in coastal sites in a particular region. Some of these sites have been contaminated by an oil spill, and some of the oiled areas have been 'cleaned', either with detergent or by scraping off the surface layers of sand or soil. Has any of this affected the beetle? The only data we have are on the presence or absence of the beetle *after* the spill, so the nearest we can come to answering this question is to see

Table 2.9   Contingency table for presence and absence of a beetle on oiled and unoiled coastal sites. Expected values in parentheses

Site type	Beetle present	Beetle absent	Total
Unoiled	80 (69.3)	10 (20.7)	90
Oiled (not cleaned)	40 (46.2)	20 (13.8)	60
Oiled and detergent-washed	39 (42.4)	16 (12.6)	55
Oiled and scraped	49 (50.1)	16 (14.9)	65
Total	208	62	270

whether the presence of the beetle is associated with site type (unoiled, oiled, detergent-washed or scraped). The data are presented as a contingency table (a table of one set of categories – presence/absence of the beetle – against another – site type) in Table 2.9.

Our best estimate of the overall presence of the beetle is 208/270, i.e. 77.04% of sites have the beetle present. If there were no influence of site type, all site types should have the beetle present in this proportion of sites. From this, we can calculate the expected number of sites of each type with the beetle present, e.g. in oiled and untreated sites, it would be $60 \times 0.7704 = 46.2$.

To see if the actual counts are significantly different from the expected ones, we calculate chi-squared, as

$$\chi^2 = \Sigma((\text{observed–expected})^2 / \text{expected}) \tag{2.6}$$

The larger the value of χ^2, the stronger the evidence for some association of beetle with site type. For Table 2.9, χ^2 is 12.06. To find the probability of this value occurring by chance, we first find the degrees of freedom (d.f.) for the contingency table. This is $(r - 1)(c - 1)$, where r and c are the number of rows and columns, in this case four rows and two columns, so d.f. = 3. Tables of χ^2 at 3 d.f. show that a value of 12.06 has a probability < 0.01, i.e. there is less than one chance in 100 of such a value occurring by chance.

From Table 2.9, we can see that the differences between the three treatments on oiled sites are quite small compared to the overall difference between oiled and unoiled sites. We might then wish to simplify the comparison, and only consider the association between the beetle and whether or not sites have been contaminated with oil, ignoring the clean-up treatment. Recombining the data in this way gives a simple 2×2 contingency table (Table 2.10). Calculating chi-squared in the same way as before, gives a value of 10.72, with $(2 - 1)(2 - 1) = 1$ d.f. This value is significant at $p < 0.01$ and close to the tabulated value of 10.83 at $p = 0.001$.

One final point to note when considering whether to use chi-squared, is that it does not work very well for very small samples. As a general rule, if more than a small proportion of expected values fall below 5, or any are below 1, the chi-squared test may not be valid. In such a case, you should use one of the many alternatives for small samples described by Siegel and Castellan (1988).

Table 2.10 Contingency table for presence and absence of a beetle on oiled and unoiled sites after an oil spill

Site type	Beetle present	Beetle absent	Total
Unoiled	80	10	90
Oiled	128	52	180
Total	208	62	270

2.5.5.2 Correlation

The above example shows how we can assess the association between groups of measurements when all we have are categorical observations. However, if we have a numerical measure for our observations, we can go a stage further, and look at the overall relationship between two variables. The simplest measure of this association is the Pearson, or product-moment, correlation (reference to 'correlation', when not otherwise specified, can usually be assumed to mean this form). It is a measure of the closeness of a linear relationship between the numerical values of two variables. The sample correlation coefficient (r) always ranges between -1 and $+1$. If the two variables are labelled as X and Y, then positive values of r suggest a tendency of X and Y to increase together. The reverse is true when r is negative; large values of X are associated with small values of Y and vice versa.

The correlation coefficient is defined as

$$r = \text{COV}\,(x, y) \,/\, \sqrt{\text{Var}(x) \times \text{Var}(y)} \tag{2.7}$$

Where COV refers to the covariance between x and y. We have already discussed the idea of variance, and how it is calculated (Chapter 1). The covariance of two variables is an analogous idea, except that instead of using the squares of a single variable (i.e. X^2 or Y^2) in the calculation, we use the product of X and Y (i.e. XY). The following example is based on a random sample of 10 birds from the starling dataset. The following are the calcium and manganese values of these birds:

	Ca	Mn
1	121.400	0.4700
2	224.900	5.4900
3	203.700	5.2800
4	169.100	4.4800
5	379.500	5.4400
6	187.000	3.4100
7	329.600	5.9800
8	440.200	6.4400
9	152.200	3.9000
10	184.500	4.4400

In this example the variance of calcium is 11310.97 and that of manganese is 2.91. The covariance is calculated as

$$\sum \frac{(x_i - \bar{x})\,(y_i - \bar{y})}{n - 1} \tag{2.8}$$

Here COV (x, y) = 133.59 and the calculated correlation coefficient is

$133.59/\sqrt{(11310.91*2.91)} = 0.736$. This exceeds the critical tabulated value of 0.632 ($p = 0.05$) and we can conclude there is evidence of correlation between calcium and manganese in this subset of data.

There are several important points to note at this stage. Firstly, correlation is a measure of numerical association. It does *not* imply that variation in one variable is the *cause* of variation in the other, only that the two are in some way linked. This might be a causal link, but it might also be a coincidental link, due to their common control by a third variable. Secondly, *r* is only a measure of *linear* correlation, that is, the degree to which one variable increases or decreases by a constant amount, for a given increment in the other. You can get a significant linear correlation when the true relationship is not linear, simply because both variables have the same overall directional trend. Some examples of this problem are given later (sections 2.6 and 2.8). Additional cautions mentioned by Clarke and Cooke (1978) are that you can obtain a spurious correlation by linking a part of one variable with the whole of another, that combining unlike populations can give problems and that you should not make any inferences about unsampled populations.

The above description is of correlation between any two variables. The concept can be extended to encompass several variables (see Chapter 6). The correlation coefficient described assumes both variables are normally distributed, but is a fairly robust measure, so will not be grossly affected as an exploratory tool except by gross deviations from normality. However, if in doubt, it is worth considering the many non-parametric measures of correlation, such as rank correlation, coefficient of concordance (a measure of association between many variables), etc. described in Siegel and Castellan.

2.6 VISUALIZING THE DATA II: COMBINATIONS OF VARIABLES

2.6.1 TABLES

The construction of summary tables is often one of the first tasks when examining datasets. Section 2.4 showed some of the ways in which a table summarizing a single variable could help in interpreting its pattern. These principles can be extended to cover the summary of more than one variable or sample in a single table.

Take for example the water quality dataset (Appendix 4). This dataset is already ordered by site and within site by age. If we look at the ages in the different sites, we see that the site code refers to a combination of study area and plantation age, and that each area, in effect, has a consistent series of ages, which we can put into age-classes, using the number included in the site code. This then enables us to construct a table of several variables in relation to age-class, to look for trends related to the growth of the trees.

Table 2.11 shows clearly four distinct patterns between age-classes. Nitrate

Table 2.11　Some water quality variables (mean ppm over all sites) in relation to plantation age-class

Age-class	NO$_3$	SO$_4$	PO$_4$	Si	pH	Ca	Mg
0	0.11	1.3	0.0004	0.64	5.5	1.4	1.0
1	0.09	1.5	0.0006	0.85	5.1	1.1	0.8
2	0.15	1.6	0.0034	0.90	4.9	1.2	0.9
3	0.49	2.0	0.0016	1.18	5.0	1.2	1.1
4	0.51	2.4	0.0000	0.95	4.9	1.6	1.3

and sulphate stay fairly steady until age-class 3, when they start to rise steeply. Phosphate and silicon show a more complex trend, rising to a peak at an intermediate age, then falling in the oldest classes. There is an apparent decrease in pH after trees are planted, and calcium and magnesium both show approximately opposite patterns to those of phosphate and silicon; they start and finish high, with lower levels at intermediate ages.

We could also use a table such as this to look for associations or correlations between the water quality variables, for example there seems to be some correlation between calcium and magnesium, or between nitrate and sulphate. Since we are concerned here with exploring patterns in the data, we do not need to test any explicit model to account for the correlation. The purpose of the table is to summarize patterns and trends in the data in a way that will help us formulate more explicit models.

We can extend this kind of table to enable multi-way comparisons. For example, Table 2.12 gives the data for silicon for each area/age combination. From this we see that although all areas have silicon rising then falling with age of trees, each area has a different pattern over time; any model we construct will have to account for both temporal and geographic variation. You might wish to examine some of the other variables in this way, to see the interaction of

Table 2.12　Silicon content of waters (annual mean), in five study areas and five plantation age-classes

Age-class	Study area (code)				
	BT	DI	DN	HN	TY
0	0.38	0.46	0.75	0.70	0.90
1	1.08	0.77	0.78	0.81	0.80
2	0.90	1.18	0.79	0.87	0.78
3	0.75	0.94	1.05	1.53	1.63
4	0.60	0.88	0.97	1.14	1.17

location and tree growth effects. A useful addition to this table might be a column of overall means for each age-class, and a row of overall means for each site, similar to the row and column totals in Table 2.10. These would show the overall patterns, against which site- and age-specific patterns could be compared.

A number of useful points helpful for constructing tables are given by Clarke and Cooke (1978) and Ehrenberg (1982).

2.6.2   OVERLAID HISTOGRAMS

Histograms of single samples were described in section 2.4.2. They can also be used to compare two or more samples if the same horizontal and vertical scales are used for all samples. If we display the Cd levels found in the male and female starlings as two separate histograms one above the other, then the two distributions can be compared (see Figure 2.7a, b). Both the positions of the 'peaks' and the shapes of the distribution are different. Figure 2.7c gives the same

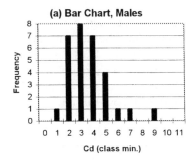

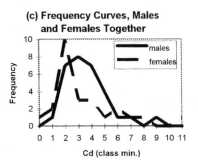

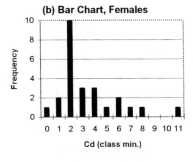

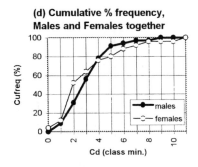

Figure 2.7   Histograms and cumulative percentage frequency curves for cadmium levels in male and female starlings. (a) and (b) Histograms, sexes separately. (c) frequency curves, sexes overlaid, (d) cumulative percentage frequency curves, sexes overlaid. The background grid and data points in (d) are to help focus on vertical or horizontal differences between curves.

data as overlaid frequency curves, and Figure 2.7d as cumulative percentage frequency curves. These are particularly useful for checking whether one distribution tends to have higher or lower values than another.

2.6.3 OVERLAID BOXPLOTS

Boxplots were explained in detail in section 2.4.3. Like histograms, boxplots of several different samples can be overlaid to show the relationships between them. Figure 2.8 shows the sulphate data from Appendix 3 within each plantation age-class. The increasing median sulphate content of the water with increasing tree age is clearly shown, but also the changes in variability can be clearly seen, in particular the very variable levels at age-class 2, and almost no variation at age-class 3. These differences in variability might be worth further investigation.

2.6.4 SCATTERPLOTS

For each of the sites recorded in the water quality data set, a number of characteristics have been recorded. If we want to examine whether any of these characteristics are correlated, we can do this by scatterplots. These are simply plots of one variable against a second. Figure 2.9 gives some examples. There is no correlation between pH and sulphate (Figure 2.9a). The relationship between pH and calcium may be one of positive correlation, but much of the

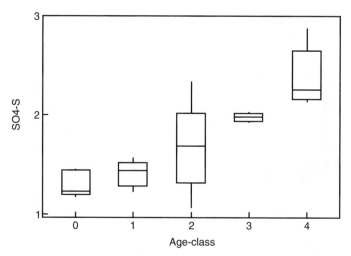

Figure 2.8   Boxplots showing changing patterns of sulphate content of waters with age of trees in plantations

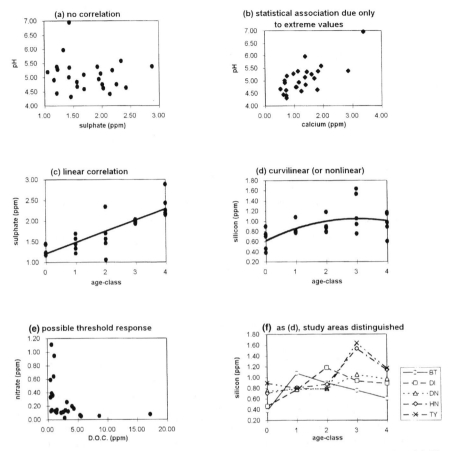

Figure 2.9 Scatterplots showing different kinds of relationships between variables. (a) No relationship; (b) apparent relationship due only to two extreme points; (c) linear correlation; (d) curved relationship; (e) threshold response; (f) as (d) with different study areas distinguished.

apparent trend is due to a single extreme point (Figure 2.9b). If that is removed, nearly all the visible association disappears. In contrast, there is a strong trend of increasing sulphate with tree age-class (Figure 2.9c). In this plot, we can also see that the wide variation at age-class 2, seen in the boxplots, is due to two relatively extreme values, which might repay a closer look. The plot for silicon (Figure 2.9d) suggests a curved trend with age; if this is plotted with different symbols for each study area, the different temporal patterns can be clearly seen (Figure 2.9f). Finally, there appears to be a 'threshold' relationship between nitrate and dissolved organic carbon (DOC) with high nitrate levels confined to DOC below about 6 (Figure 2.9e). However, we know from the silicon plot that in some cases the age-related trend is partly because of differences between study areas. You might like to try plotting these two variables distinguishing between areas, to see if this is so here too.

2.6.5 MORE COMPLEX VISUALIZATION

This area of exploration consists of topics such as response surfaces, three-dimensional plots, contours and use of multiple axes/scales to name but a few of the many techniques that the modern computer graphics and statistical package has available. Figure 2.10 gives a few examples. Please read the small print before you use them as they are excellent at producing very fancy diagrams from very limited data sets. They may look good, but they may also be effectively meaningless. Compare, for example, Figure 2.10a with b and c. Smoothed interpolation curves as in figure 2.10d can be a very useful exploratory tool, especially when combined with the original data points as a way of seeing the *shape* of a relationship. This may then suggest more formal models against which the data can be tested. Again, however, please remember that in this exploratory view, you are not testing any explicit model, only investigating possible shapes.

2.7 DIVERSITY/RICHNESS INDICES

In section 2.3 we described some summary measures for single variables, but what if your data are, for example, the species composition of the vegetation at a site, or measures of the absolute or relative amounts of different habitat elements? Such data might be gathered, for example, where the effect of a low-level pollutant could be expected to affect many different species in different ways, with perhaps no consistent effect on all species, but changing the balance of species. Excess nitrogen applied to grasslands, for example, could give one or more species a competitive advantage over most others, so that they became more dominant in the vegetation. In that situation, we might feel that the over-all species diversity of the grassland had decreased, and richness and diversity indices are one way of expressing measurements of several related variables (such as species in a biological community) in a single summary measure. These indices can thus be thought of as a multivariate summary measure, akin to the measures of variation or shape for single variables described in section 2.3.

Before going into individual indices, it is probably worth clarifying what we mean by 'richness' and 'diversity'. The two terms tend to be used loosely, and sometimes interchangeably, but they each deal with a rather different aspect of the variation among the constituent items in a population. So, 'species richness' indices usually measure the number of species in a given sampling unit, but a 'rich' habitat may refer to the abundance of individuals supported by that habitat, with no necessary connection to the number of *different species* supported.

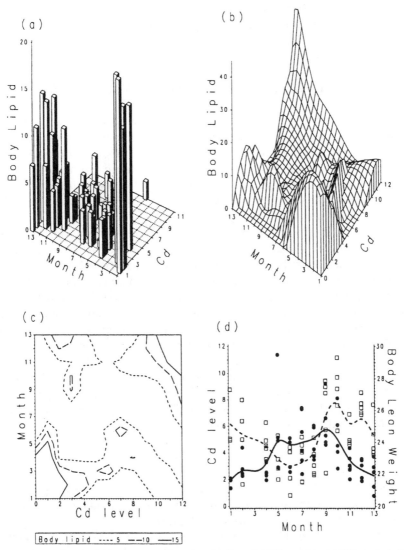

Figure 2.10 Examples of more complex visualisations (a) 3D scatter (b) 3D surface interpolated from the data in (a) – note the entirely spurious peaks at the back (c) contour plot of
interpolated data used in (b) – alternative view, interpolation has produced some spurious
peaks – note contours may show features that are hidden in a 3D surface view (d) multiple
axes/scales and smoothed interpolation curves (can be useful to get an idea about shape of
response, should not be used for explicit modelling).

Similarly, while 'diversity' indices usually measure the variation among different
categories (e.g. relative proportions of species in a community), the literature
abounds with the use of 'species richness' as an indicator of 'biodiversity'. For
our purposes here, we adopt the general distinction that 'diversity' is primarily

concerned with the variety of different categories, while 'richness' gives more emphasis to the actual amounts of each or all categories (see French 1994, for further discussion of this point).

2.7.1   MAIN TYPES OF RICHNESS AND DIVERSITY MEASURES

There are an enormous number of different diversity and richness measures; we will describe in detail only a few. However, this plethora of measures mostly fall into three main categories: species richness indices, proportional abundance indices, and species-abundance (frequency distribution) models. The last of these, however, are more complex frequency models, rather than simple single indices, so outside the scope of this section. For some discussion of their use and its problems, see Wilson (1991 and 1993).

Of the other two types, *species richness indices* only give information on the number of categories present. The most commonly used is the index usually referred to as species richness, which is simply the number of different species present in a sample. This is easily understood, and very simple to calculate. However, it is not adequate if we are at all interested in how much of each species there is – or how many individuals in the whole sample. For example, consider two samples of soil invertebrates. Both contain 10 species, so both have the same species richness index. But what if one has fewer than five individuals of each species and the other over 50 of each? They are clearly not the same, but the simple species richness index cannot detect the difference.

Proportional abundance indices fall largely into two groups: information theory indices, and dominance indices. The *Shannon index* (H), an information theory index, is defined mathematically as

$$H = -\sum_{i=1}^{n} (p_i \ln p_i) \tag{2.9}$$

where n is the number of species (or other categories), and p_i is the proportion of the ith species in the total. The maximum value H can take depends on the total number of species in the sample, and is $\ln(n)$.

A simple example will illustrate the procedure. Imagine a small woodland contains 10 bird species. The numbers of territories held by each species are 5,5,4,4,3,2,2,1,1 and 1. The total number of territories is thus 30, and H, calculated as above, is 2.07. The possible maximum value for 10 species is 2.30; the high value of H in this example is, because the total number of territories is shared among many species, and there are no clearly dominant species. If the first species had been very dominant, say having 21 territories, with the rest all having only one, H would be much lower at 1.27.

Strictly, the above formula only applies to a complete census of a population or community. If H is estimated from a sample, the formula given produces a biased value for H. In practice, the error is rarely of any significant size and the census version is commonly used. A more substantial potential error comes from the fact that samples of most communities miss some species, and this error increases as the sample size decreases relative to the whole community. The size of this second error is usually unknown, but its probable presence must always be remembered when interpreting Shannon indices.

Simpson's index (*D*), is a commonly used dominance index, calculated as the sum of squared proportions of each species in the population. i.e.

$$D = \Sigma p_i{}^2 \tag{2.10}$$

where p_i is the proportion of the *i*th species. This formula gives increasing values of *D* with increasing dominance of one or a few species, i.e. decreasing values of *D* with increasing diversity. When used as a measure of diversity, it is therefore usually expressed as its reciprocal (1/*D*) or as $1 - D$.

As with the Shannon index, the formula above is only strictly valid for an infinite population. A finite-sample version exists but has several drawbacks, discussed by French (1994).

Of the above types of measure, species richness indices are true measures of 'richness', while proportional abundance indices are measures of 'diversity' in the strict sense, i.e. they measure the partitioning of the overall species variation among the different species, irrespective of any 'richness' element. In practice, however, we are frequently interested in the *combination* of richness *and* diversity. One index which explicitly combines both richness and diversity elements in a single measure is the hierarchical richness index (HRI). The data scores are assigned to groups (e.g. abundance scores to species, or species to functional groups e.g. by habitat affinities) and the total abundance score calculated for each group. The groups are then ranked in descending order of abundance score (i.e. highest score ranks 1, next highest 2, etc.) and HRI calculated as

$$HRI = \sum_{i=1}^{g} (s_i \times i) \tag{2.11}$$

where *g* is the number of groups, *i* the group rank and s_i the within-group abundance measure for the *i*th group.

A simple example illustrates the relationship between species richness indices, proportional abundance indices and HRI. Suppose we were investigating effects of changing pesticide application on small mammals in farm fields. Two fields form part of two different treatments. Both contain five species of mice, voles, etc. but one field has three individuals of each species, the other has 5, 8, 10, 20 and 50. Their species richness scores are identical. This is not, for our purposes, very useful. Proportional abundance indices would score the first field much higher than the second. This is, in this case, a misleading result, since if we ask which treatment is 'better' for small mammals, the answer is obviously the second. HRI would give the 'correct' answer in this case. However, if it was more important to have an even balance of species, a proportional abundance index would fit our purpose better, while if we were concerned only with keeping all species at least present, neither proportional abundance indices nor HRI would have any advantage over simple species richness. As usual, it is important to choose a suitable measure to fit the question of interest. For more detailed discussion of these aspects, see French (1994).

2.7.2   USE AND INTERPRETATION OF RICHNESS/DIVERSITY INDICES

Whatever measures of diversity or richness may be of interest in any particular study, two general points must be emphasized. First, no simple index can give a comprehensive picture of the 'richness' or 'diversity' within a sample. Still less can any such index provide a full assessment of any differences between samples. Species richness emphasizes the variety of species, giving all species equal weight, hence gives greater weight to rare species (and correspondingly less to common ones) compared with proportional abundance indices. Neither species richness nor proportional abundance indices take into account the total abundance of all 'species' in the sample; HRI does this, but has its own drawbacks. This problem can be partly overcome, particularly in exploratory analyses, by calculating several different indices on the same data, and seeing if they all tell the same story. Any major differences between the picture given by different indices can also often be interpreted in terms of the ecological processes or patterns involved.

Secondly, richness or diversity indices are frequently used to assess the 'conservation' value' of a site of community, or the effects on it of some external factor such as pollutants, toxic emissions or changes in management of the ecosystem. The uncritical use of aggregate numerical indices of the type described here can then lead to serious errors in ecological interpretation, since they take no account of any qualitative features. For example, a site might gain in its numerical richness or diversity score after invasion by common opportunistic species, at the expense of more specialized or 'typical' species of greater 'conservation value' (on whatever criteria that might be decided). There would thus be an overall *loss* of richness, in some sense, from the original habitat, even though its richness or diversity, as measured by a standard index, had increased.

Finally, there is the general problem of the physical size of a sample unit in relation to the scale of pattern in the system being studied. There might, for example, be a change in the spatial distribution of species, so that in a small area there were more species than before, but the habitat was also becoming more homogeneous, so over a larger area many relatively rare species, formerly living in small specialised microhabitats, were being lost. Here, you would get conflicting answers, depending on the size of your sample plots; small plots would give results suggesting an increase in species richness/diversity, large

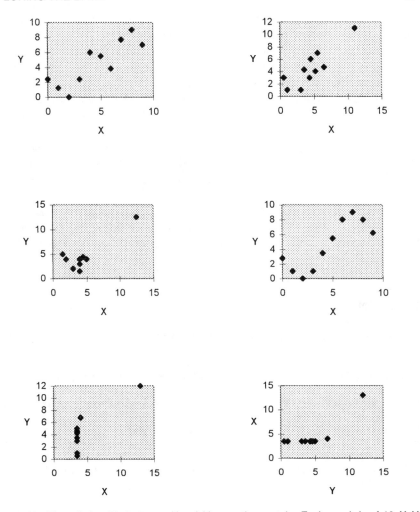

Figure 2.11 The relationship between *X* and *Y*: a cautionary tale. Each graph is of 10 *X–Y* points. Every *X* has mean = 4.5 and standard deviation = 3. Every *Y* has mean = 4.5 and standard deviation = 3. Every *X–Y* pair has linear (correlation (r) = 0.84 ($p < 0.01$).

plots would give results suggesting a decrease. Which one would be 'right'?

2.8   VISUALIZING THE DATA III: A CAUTIONARY TALE

We have in this chapter described various techniques for exploring your data. Among these are various kinds of visualisation, and we always recommend a graphical examination of your data – 'If in doubt, plot it out.' The value of that can be illustrated by considering Figure 2.11.

Here are six scatterplots of variables X and Y. Each consists of 10 points and each shows a different pattern – a simple linear trend, a possible linear trend, but 'stretched' a bit by one extreme point, a clump of points with a single outlier and otherwise no relationship between X and Y, a strongly curvilinear relationship, and one which might be asymptotic, or if we were interested in X in relation to Y might indicate a threshold response. Yet in all of these datasets, the mean value of both X and Y is 4.5, the standard deviation of both variables is 3, and the linear correlation (r) between them is *in every case* 0.84 – a value having a probability under the null hypothesis of $p < 0.01$, which would in most cases be taken to be highly significant, but in all but one of these five cases would probably *not* be a correct expression of the true relationship between X and Y.

Uncritical application of statistical tests without preliminary data exploration would in these cases have produced completely misleading results – but use of the exploratory techniques outlined in this chapter will greatly reduce the risk of such mistakes.

REFERENCES

Clark GM and Cooke D (1978) *A Basic Course in Statistics*, Edward Arnold, London.

Ehrenberg, ASC (1982) *A Primer in Data Reduction: An Introductory Statistics Textbook*, Wiley, Chichester.

French DD (1994) Hierarchical Richness Index (HRI): a simple procedure for scoring 'richness', for use with grouped data. *Biological Conservation*, **69** 207–212.

Siegel S and Castellan NJ (1988) *Nonparametric Statistics for the Behavioural Sciences*, McGraw-Hill, New York.

Sokal RR and Rohlf FJ (1995) *Biometry*, Freeman, New York.

Wilson JB (1991) Methods for fitting dominance/diversity curves. *Journal of Vegetation Science*, **2** 35–46.

Wilson JB (1993) Would we recognise a broken-stick community if we found one? *Oikos*, **67** 181–183.

3

Field Experiments

DICK GADSDEN[1] AND TIM SPARKS[2]

[1]*School of Computing and Management Sciences, Sheffield Hallam University, UK*
[2]*Institute of Terrestrial Ecology, Monks Wood, UK*

3.1 INTRODUCTION

Historically, many statistical ideas of experimental design have developed hand-in-hand with agricultural field experimentation. The need to obtain reliable data and to carry out justifiable analysis of that data has led to many aspects of good practice being generated. In this process of design there are responsibilities on both the experimenter and the statistician. The experimenter should define what it is he/she wishes to investigate in clear terms, including all comparisons that need to be carried out. An appropriate experimental design can then be planned in consultation with the statistician and then applied to the problem. This design should be chosen to be as efficient and effective as possible, that is to yield maximum information for minimum effort. This does not mean minimal data collection, but collecting enough data to allow appropriate comparisons to be made. When collected these data can then be analysed statistically and appropriate conclusions drawn.

When designing the experiment, as well as the context of the problem itself, there are three principles which should be borne in mind. These are the concepts of randomization, replication and local control. To start with, though, let us consider what happens when we make an experimental observation. We hope and expect that it will reflect an unbiased measurement of some effect we are interested in. Suppose we have applied some contaminant at various doses to field plots, then the observed concentration in earthworms where a particular dose was applied might be described as

$$
\text{observed concentration} = \begin{array}{c} \text{concentration in the absence of dosing} + \\ \text{concentration due to dose} \end{array} \quad (3.1)
$$

However, if we were to repeat this process we would not expect both observed concentrations to be exactly the same. Rather, we know that natural or random variation will usually lead to different observation values. This is known as

Statistics in Ecotoxicology. Edited by T. Sparks. © 2000 John Wiley & Sons Ltd

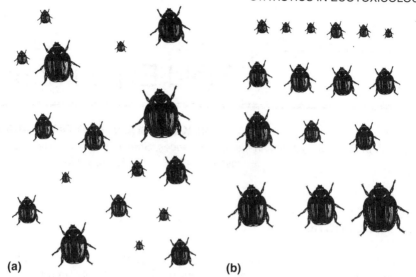

(a) **(b)**

Figure 3.1 (a) a hypothetical example showing extreme variation in beetle size (b) same beetles but rearranged, in rows, according to some feature of sampling which seems to account for much of the variation in size, although within-row variation is still evident.

residual variation for experimental observations. To take account of this we can redefine the above description as

$$
\text{observed concentration} = \text{concentration in the absence of dosing} + \\
\text{concentration due to dose} + \qquad (3.2) \\
\text{residual variation}
$$

The first two parts of this decomposition are known as *explained variation* and the last part as *unexplained variation*. The aim of designing statistical experiments is to ensure all effects which can be explained are taken into account and as little unexplained variation as possible is left.

As a graphical representation of this consider a sample of beetles where there is considerable size variation (Figure 3.1a). Some of the variability in size can be attributed to some feature of the sampling (such as origin or species), but even when they are presented in separate categories (Figure 3.1b) it is still possible to see that there is considerable unexplained variation present.

3.2 TERMINOLOGY

Most of the terminology used in experimental design derives from its development within an agricultural context.

An *experimental unit* is the physical item which is experimented upon, usually that on which 'treatments' are applied.

This means that in the example where a contaminant was applied to an area of land, that area of land is the experimental unit. In other applications it could be a plant, animal or any physical entity.

> A *treatment* is a characteristic of how the experimental unit has been modified.

Again, in the above example, the contaminant used is the treatment. If we are comparing different contaminants then each would be a different treatment. Treatments can also be a characteristic of the experimental unit itself that we are interested in investigating. In a trial concerning birds where we wish to see whether male and female birds lead to different results, the two sexes constitute two different treatments.

> A *factor* is a series of treatments which are collectively to be compared; treatments may be considered to be the levels of the factor.

Hence, the collection of different contaminants constitute a factor. Similarly, gender will constitute a factor in the bird investigation.

> *Example 3.1*
> Suppose a field is divided into eight plots. An experiment is to be carried out to compare four contaminants and two land-management techniques. Each contaminant/land-management combination is to be applied to one plot.

Here, there are eight units, the plots of land. There are two factors, the contaminants and the land-management techniques. Within one factor there are four treatments (the contaminants) and in the other factor two treatments (the land-management techniques).

3.2.1 RANDOMIZATION

When designing an experiment it is important to ensure that no bias, whether intentional or unintentional, is introduced into the results. Random allocation of treatments to experimental units is essential to ensure that observed results are unbiased and are not confounded with other known or unknown factors. Where randomization is found to be impractical then the likely consequences should be carefully evaluated and discussed in relation to any conclusions drawn.

3.2.2 REPLICATION

Replication is the repeating of treatment combinations on more than one experimental unit. It allows uncontrollable unit-to-unit or residual variation to be estimated in an experiment and this can then be used as the basis for comparison of treatment effects. The accuracy of an experiment can be improved through replication.

Replication in the statistical sense refers to independent replicates and not to a subsampling of a single experimental unit. Where an experimenter carries out subsampling it is known as pseudoreplication (see also section 3.7.2) and this generally leads to an underestimate of residual variation and hence to potentially incorrect conclusions when comparisons are made. It also draws into question what might have caused any difference in results, is it the treatments themselves or a 'location' effect due to the pseudoreplicates being from the same 'location' or original unit?

Example 3.2
In the experiment in Example 3.1, pseudoreplication would occur if, having allocated treatment combinations to areas, those areas were subdivided to obtain more than one result. To ensure independent replicates, the areas should be subdivided first and then treatment combinations allocated randomly to the resulting subdivided areas.

3.2.3   LOCAL CONTROL

In many investigations we may know that a particular factor is likely to influence our results but we do not actually wish to investigate that factor, i.e. it is just a nuisance. In designing our experiment we would try to ensure this factor had equal influence on the factor(s) that we do wish to investigate.

Blocking is the allocation of experimental units to discrete groups such that the units are as equivalent as possible in relation to an extraneous factor.

Block effects can occur in a number of ways. It may be that soil conditions are not uniform within a field. It might be that only a certain number of observations can be made at a certain time. It might be that observations are being made in different parts of the country.

Example 3.3
Suppose a field has three types of soil within its boundary: clay, sandy and shale. If our experiment is not concerned with investigating soil types, but only the allocated treatments, then we would need to ensure our experimental design allocated each of the treatments to the soil types, so that the soil type effect is as equal as possible on the observed results.

A design like Figure 3.2 would not be acceptable since we would not be able to tell whether any differences in results were due to the treatments or the soil

CLAY	A	A	A
SAND	B	B	B
SHALE	C	C	C

Figure 3.2

CLAY	B	A	C
SAND	C	A	B
SHALE	A	B	C

Figure 3.3

type (A goes with clay, etc.) However, the design in Figure 3.3 ensures that each treatment type is used within each soil type.

> *Confounding* in experimental design is where the effect of one factor cannot be separated from the effect of another factor.

In the design in Figure 3.2 confounding has occurred between soil type and treatments. This is relatively easyto identify, but in any design it is important to consider whether an unconsidered or latent factor might cause unexpected confounding and hence make it difficult or impossible to attribute the real cause of an observed difference in results. Partial confounding can also cloud interpretation of results.

> A *balanced* design is one where every treatment combination appears an equal number of times within a block.

In the design in Figure 3.3 there is only one factor, and each treatment level (A, B, C) appears once in each block (soil type). Thus, this is a balanced design.

3.3 FIELD EXPERIMENTS IN ECOTOXICOLOGY

The aim of field experimentation is to evaluate the differences or effects of selected treatments in real-life situations. Field experiments are inherently more variable than laboratory studies (see Chapter 4) although steps are taken to eliminate or minimize random and nuisance variation.

Typical ecotoxicological experiments involve the contamination of replicated experimental units and a comparison between different types of contamination, between different doses of the contaminant or between contaminated and control (zero dose) treatments. Steps may need to be taken to ensure that there is no exchange of contaminant between adjacent plots and this can be achieved by ensuring adequate spacing between plots (guards) or by containing contamination within, for example, mesocosms.

In agriculture, experiments assessing typical responses such as yield or quality are very common and the levels of likely variability broadly known. Ecotoxicological experiments, in contrast, can be considered to be less uniform in goals and measurements and the measurements themselves can be considerably more variable. Thus it is often necessary to undertake a greater intensity of recording and/or replication to achieve the same level of precision (see also section 3.7.2).

3.4 DESIGNS

We have introduced many of the ideas which should be considered when designing an experiment. A number of standard designs have been generated in light of these considerations and some of these are outlined in this section.

Before developing individual designs there are a number of considerations with regard to assumptions about the data collected. In many planned experimental situations there is a tendency to apply analysis of variance (ANOVA) techniques to data collected without considering the nature of the data itself. While this may be appropriate in many practical situations, this approach involves implicit assumptions. To apply ANOVA we are assuming that the unexplained or residual part of an observation has expected value zero, i.e. tends to cancel out with other observed residuals. We are also assuming that one residual value has no effect on any other residual value, i.e. all residuals are independent of one another. Further, we are assuming that all residual values have the same underlying variability. Finally, to allow us to carry out statistical tests, we assume the residuals are normally distributed.

These assumptions are reasonable for many physical measurements, but must be borne in mind for each and every specific problem before applying ANOVA techniques. Sometimes these assumptions can be more nearly met once data have been transformed.

In designing experiments, factors being investigated can be different in nature. While the method of analysis is essentially unaltered, the type of factor being investigated affects the conclusions that can be drawn. Two types of factors are generally considered, *fixed* and *random* factors.

A *fixed* factor is a factor where the treatments being compared are the only ones to be investigated or for which conclusions are to be drawn.

In Example 3.1 with four contaminants, this would be a fixed factor if these were the only four contaminants we were interested in or wished to draw conclusions about. The four contaminants in Example 3.1 would constitute a random factor if we wished to draw conclusions about a wider group of contaminants, but were only able to investigate this set of four contaminants as representative of all the contaminants of interest. They would be chosen from the full possible group of contaminants by some process which ensured they were representative of the full group, often by a random process.

A *random* factor is a factor where the treatments being compared are representative of a wider collection of treatments. Conclusions are to be drawn about a wider range of treatments.

3.4.1 COMPLETELY RANDOMIZED DESIGN

In a completely randomized design all treatment combinations are allocated to experimental units by a random process. Equal or unequal numbers of replicates per treatment are possible.

Example 3.4
Suppose an experiment is designed where four contaminants are to be compared for their effect on herbage yield. The experiment is to be carried out in a field which has been divided into 16 equal-area plots. Design a completely randomized experiment for this situation.

Solution 3.4
Here, four plots can be allocated to each of the contaminants giving four replicates. If we categorize the contaminants as A, B, C, D we need to allocate these to the 16 plots by some random process so that each is repeated four times. A possible random allocation is

A	C	A	B	C	B	C	D
D	B	B	D	D	A	A	C

Usually, random numbers or drawing of lots are used to make this allocation.

To analyse this design a one-way ANOVA is used.

Example 3.5
In the above example, suppose the observed herbage yields from the equal area plots were observed to be

56	48	52	46	51	44	49	48
47	43	46	50	52	49	53	51

then the results can be rearranged as

	Contaminants			
	A	B	C	D
Replicates	56	43	48	47
	52	46	51	50
	49	46	49	52
	53	44	51	48

and the resulting analysis of variance table is

Source	Degrees of freedom (d.f.)	Sums of squares (ss)	Mean square (ms)	F	Prob $> F$
Contaminants	3	123.69	41.23	9.29	0.0019
Residual	12	53.25	4.44		
Total	15	176.94			

Here, the F-test tells us that the factor 'contaminants' is significant, in fact highly significant. There are some contaminants that differ from others in respect of their effect on herbage yield. It does not tell us which differ from which, only that there are differences. This conclusion is drawn on the basis of how likely the observed result would occur from random sampling of equivalent treatments. Here, the very small p-value indicates this is an unlikely event to occur by chance, and we conclude that there is a statistically significant difference between contaminants.

If in our example the four contaminants are the only ones of interest (i.e. a fixed factor) we can say there are differences between contaminants A, B, C, D. However, if the four contaminants are only representative of a wider group of contaminants (i.e. a random factor) we can only say that contaminants in the wider group, in general, differ in their effect on herbage yield.

When we have a fixed factor we can carry out further analysis to ascertain which specific contaminants differ from which (see sections 3.5 and 3.6). This would not make sense for a random factor since we are drawing conclusions about a wider group, not how representative contaminants might or might not vary relative to each other.

3.4.2 RANDOMIZED COMPLETE BLOCK DESIGN

In a randomized complete block design the treatment combinations to be compared in the experiment are repeated an equal number of times (usually once) within a block. The treatment combinations are allocated randomly to the experimental units within the block. The word 'complete' is used to indicate the full set of treatment combinations has been repeated an equal number of times within the block.

Example 3.6
Suppose an experiment is to be designed where four contaminants are to be compared for their effect on herbage yield. The experiment is to be carried out in two fields where sandy and clay soils respectively are present. The experimenter is interested in differences in yield due to the contaminants but not the soil conditions.
   Design a randomized complete block experiment for this situation.

Solution 3.6
At least four plots or experimental units are required within each of the soil type areas. If we categorize the contaminants as A, B, C, D we need to allocate the contaminants A, B, C, D to the four plots within each of the soil types. A possible random allocation is

Sandy

A	B	D	C

Clay

B	A	C	D

Here, we have a randomized complete block design with single replicates in each block. To analyse a randomized complete block design we carry out an ANOVA. If the blocking factor has an effect then we have been justified in using a block design. These randomized complete block designs are the most common design in field experiments and have many logistical advantages over completely randomized designs.

Example 3.7
In the above example suppose the observed herbage yields from equal area plots were observed to be

Sandy

37	29	55	37

Clay

28	34	34	50

Then the results can be rearranged as

	Soil type	Sandy	Clay
	A	37	34
Contaminant	B	29	28
	C	37	34
	D	55	50

and the resulting analysis of variance table is

Source	d.f	ss	ms	F	Prob $> F$
Soils (blocks)	1	18.00	18.00	13.50	0.0349
Contaminants	3	626.00	208.67	156.50	0.0009
Residual	3	4.00	1.33		
Total	7	648.00			

Here, the blocks effect is significant so the analysis is reasonable and the design has been effective. This then means that, as contaminants are statistically highly significant, we have evidence of a difference between the effects of some of the four contaminants (fixed factors are assumed).

3.4.3 FACTORIAL DESIGNS

So far all the designs covered have included only one factor of interest. Where more than one factor is to be compared simultaneously the experiment is termed a factorial experiment and an appropriate design is required.

The simplest design is the extension of the completely randomized design.

Example 3.8
Suppose an experiment is to be designed where four contaminants are to be compared for their effect on herbage yield. In addition, two land management techniques are to be used and their effects considered. Eight plots of equal area are available for the experiment. Design an experiment to investigate both the effect of the contaminants and the land-management techniques.

Solution 3.8
Remembering the randomization principle, each combination of treatments should be allocated to one plot by a random process. A possible random allocation is

A2	B2	A1	D1
B1	D2	C1	C2

if the contaminants are denoted by A, B, C, D respectively and the land-management techniques by 1, 2 respectively.

As there is only one observation for each combination this is a factorial design without replication. When analysed we are only able to investigate the factors separately.

Example 3.9

In the above example suppose the observed herbage yields from the equal area plots were observed to be

67	73	79	80
86	68	77	68

then the results can be rearranged as

Land management technique

		1	2
	A	79	67
Contaminant	B	86	73
	C	80	68
	D	77	68

and the resulting analysis of variance table is

Source	d.f.	ss	ms	F	Prob $> F$
Contaminants	3	62.50	20.83	13.89	0.0289
Land management	1	264.50	264.50	176.33	0.0009
Residual	3	4.50	1.50		
Total	7	331.50			

Both factors are statistically significant so again, if these are fixed factors, we are able to conclude that some of the four contaminants differ from each other and that the two land-management techniques differ from each other in their effect on yield. The contaminants are significant and the land-management techniques very significant.

In this design it was not possible to investigate combinations of treatments (interactions) from the two factors. To be able to carry out such an investigation replication of the design is necessary. This does not mean a repeat of the design is used, but rather a design where each treatment combination is replicated at least twice.

Example 3.10

For the above example formulate a replicate design if 16 equal-area plots are available.

Solution 3.10
In this case a possible random allocation is

C2	C1	A1	B2
D1	A2	B2	C2
B1	A1	D2	B1
D2	D1	C1	A2

Notice there is no attempt to ensure each treatment appears once and only once in each row or column. The assumption is all 16 plots or experimental units are homogeneous. If we believed there might be row and/or column effects then we would need to build this into the design in the form of blocking.

Example 3.11
In Example 3.10 suppose the observed herbage yields from the equal area plots were observed to be

64	60	73	63
59	70	62	68
71	64	59	62
55	58	59	59

then the results can be rearranged as

		Land-management technique	
		1	2
	A	73 64	70 59
Contaminant	B	71 62	63 62
	C	60 59	64 68
	D	59 58	59 55

and the resulting analysis of variance table is

Source	d.f	ss	ms	F	Prob > F
Contaminants	3	168.25	56.08	2.82	0.1070
Land management	1	2.25	2.25	0.11	0.7452
Interaction	3	74.25	24.75	1.25	0.3558
Residual	8	159.00	19.88		
Total	15	403.75			

In this example none of the statistical tests are statistically significant. This tells us that there is no statistical difference between the effects of the contaminants, the land-management techniques or the interaction between the two.

Here the interaction is the effect due to treatment combinations across the factors and is often denoted by contaminants $\times$ land management, the two factors involved. If this is found to be significant it suggests one or more treatment combination behaves out of pattern from that expected if each treatment effect is taken into account. Such unusual results are often of major importance when carrying out an investigation.

Factorial designs can involve as many factors with as many factor levels as we wish. However, the number of experimental units required rises very quickly as the number of factors and/or levels increases, so care needs to be taken to choose only those factors and levels of factor which are really relevant to the investigation being carried out. Typically, excessive experimental ambition is diluted by a realization of the amount of effort and expense necessary to undertake the work.

In Example 3.12, three factors are involved. All the combinations of factors lead to separate interactions, each of which can then be analysed. In this analysis there are no statistically different main effects (individual factors) or interactions. In fact as the probability levels are so high in a number of cases this might suggest the analysis of variance assumptions have been violated.

Example 3.12

A comparison is to be made of how various conditions affect the growth of a variety of oak. Thirty-two oak saplings raised under similar conditions and having achieved similar growth are available as experimental units. They are to be laid out in a 4×8 pattern, where the saplings are to be well spaced. The experiment is to investigate the effects of low and high watering, addition (or not) of cadmium and whether the soil where the oak sapling is planted has been prepared or not by thorough digging.

The experimental design used was a three-factor design with four replicates. Using a code of xyz, where $x = 0$ for low watering, $x = 1$ for high watering, $y = 0$ for no cadmium, $y = 1$ for added cadmium, $z = 0$ for unprepared soil, $z = 1$ for prepared soil, the random allocation was

010	100	001	011	101	000	001	000
001	110	000	111	010	110	101	100
111	011	101	001	100	011	111	010
100	111	110	101	000	011	110	010

The results obtained were (in centimetres during six months)

32.7	32.5	30.7	37.5	26.8	29.4	26.6	31.6
28.5	36.4	29.9	29.5	23.5	27.8	37.5	33.2
31.9	28.3	30.5	26.8	27.9	25.6	36.8	34.0
27.5	26.9	30.4	30.1	30.7	25.8	35.8	29.0

Sorting these into results by each factor combination we have

		Unprepared soil ($z = 0$)	Prepared soil ($z = 1$)
No cadmium ($y = 0$)	Low watering ($x = 0$)	29.4 31.6 29.9 30.7	30.7 26.6 28.5 26.8
	High watering ($x = 1$)	32.5 33.2 27.9 27.5	26.8 37.5 30.5 30.1
Cadmium ($y = 1$)	Low watering ($x = 0$)	32.7 23.5 34.0 29.0	37.5 28.3 25.6 25.8
	High watering ($x = 1$)	36.4 27.8 30.4 35.8	29.5 31.9 36.8 26.9

and the resulting analysis of variance is

Source	d.f	ss	ms	F	Prob > F
Watering (W)	1	29.64	29.64	1.93	0.1774
Cadmium (C)	1	4.20	4.20	0.27	0.6056
Soil (S)	1	4.80	4.80	0.31	0.5811
W × C	1	1.62	1.62	0.11	0.7481
W × S	1	2.88	2.88	0.19	0.6688
C × S	1	0.12	0.12	0.01	0.9289
W × C × S	1	8.00	8.00	0.52	0.4774
Residual	24	368.52	15.36		
Total	31	419.80			

Although it is not significant in this case if a third-order or higher interaction (e.g. W × C × S) is significant it can be very difficult, if not impossible, to interpret what this represents.

3.4.4. LATIN SQUARE DESIGN

A Latin square design is a special design which allows three factors (with the same number of levels) to be analysed simultaneously. No interaction effects can be considered in the analysis and so it is necessary to assume these types of effect do not exist in the particular problem under investigation. The layout of the experiment can be represented by a square with the rows representing the levels of one factor, the columns representing the levels of a second factor and finally the third factor levels being allocated for each row and column element.

Example 3.13
A 3 × 3 Latin square might be

		Factor II		
		1	2	3
Factor I	1	A	B	C
	2	B	C	A
	3	C	A	B

Notice how the design is balanced so that each level of the third factor (A, B, C) appears once and only once in each row, once and only once in each column. To ensure the design is randomized the factor levels should be allocated randomly to 1, 2, 3 of factor I, 1, 2, 3 of factor II and A, B, C of factor III. An alternative is to allocate the rows and then the columns of the square at random. In field trials, the rows and columns may represent the physical layout in the field, i.e. both horizontal and vertical blocking.

Example 3.14
To investigate the growth of a type of moss in different conditions, four levels of watering are to be used, four different types of soil used and four levels of nitrate additive applied.

Growing tubes are set out in a square pattern with each row being a different level of watering (none, 10 ml per day, 20 ml per day, 30 ml per day), each column being a different level of nitrate additive (none, 5 ml per day, 10 ml per day, 15 ml per day) and the soils used (A, B, C, D) according to the following pattern (after randomization).

		Nitrate additive per day			
		0	5	10	15
Water	0	A	D	B	C
Per day	10	D	C	A	B
	20	B	A	C	D
	30	C	B	D	A

The results obtained were

89.3	93.2	94.1	94.3
106.4	123.6	126.4	125.8
102.3	117.5	122.4	123.6
101.1	112.4	114.6	119.2

and the resulting analysis of variance table is

Source	d.f	ss	ms	F	Prob > F
Water	3	1816.24	605.41	49.61	0.0001
Nitrate	3	634.69	211.56	17.34	0.0023
Soil	3	45.03	15.01	1.23	0.3778
Residual	6	73.22	12.20		
Total	15	2569.18			

Here, the different levels of watering have a very highly significant effect on growth as do the different levels of nitrate additive. The soil types, however, do not have any effect.

Latin square designs are not as limited as they seem. They are one of a family of row-and-column designs where the restriction is that the number of rows and columns is a multiple of the number of treatments.

3.4.5 NESTED DESIGNS

So far all the designs described have been crossed designs since each combination of factor level has been able to be allocated and observations made. Consider a situation where we wish to investigate whether three different regions of the UK lead to different responses. In each region under investigation suppose there are two soil types, sandy and rocky, which form part of the investigation. The first (or main) factor is the region. Soil type is then a second or subfactor within a region since the sandy soil in one region is not the same sandy soil as in another region, etc. The design might be drawn as

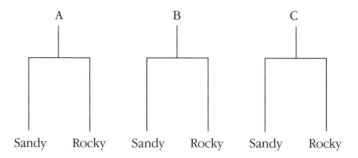

and we might then have replicates by choosing random plots within each region and soil type.

This type of design is known as an hierarchical or nested design. Because of the nature of the data collected, the interaction term that we saw in previous crossed designs does not exist. This is because we do not have the equivalent of each region combined with each soil type from a homogeneous area of soil. Rather, we have each soil type from within a region and our analysis needs to reflect this.

Example 3.15
Suppose in the three regions, two random plots are chosen from within each soil type area, leading to the following results. Carry out an analysis of these responses from equal area plots.

			Region		
			A	B	C
Soil type		Sandy	93 81	100 87	99 95
		Rocky	75 68	76 82	82 69

Solution 3.15
If a crossed design were *incorrectly* assumed then the analysis of variance table would be

Source	d.f.	ss	ms	F	Prob > F
Regions	2	130.67	65.33	1.34	0.3292
Soil	1	884.08	884.08	18.20	0.0053
Regions × soil	2	28.67	14.33	0.30	0.7547
Residual	6	291.50	48.58		
Total	11	1334.92			

However, because soil type is nested within regions, the soil type and interaction term should be combined to give a 'soil type within regions' factor. The correct analysis, where 'regions' are tested against 'soils (within regions)', is

Source	d.f	ss	ms	F	Prob > F
Regions	2	130.67	65.33	0.07	0.9339
Soils (within regions)	3	912.75	304.25	6.26	0.0281
Residual	6	291.50	48.58		
Total	11	1334.92			

In this analysis the main effect for soils and the interaction are combined into the *within* term. Notice that in the correct analysis the difference in effects from the soils is a lot less significant although still statistically significant.

3.4.6   SPLIT PLOT DESIGNS

A split-plot design is an extension of nested designs where we are unable to fully randomize our treatment allocations. Suppose an experiment has been carried out as follows:

Example 3.16
Three fields have been divided into two, one half being left untreated (N), the other half having been treated with a soil insecticide (Y). Four varieties of potato (A, B, C, D) have then been allocated randomly to four equal areas within each half field, giving the following overall design (a suitable gap has been left between the plots in each field):

Field

1		2		3	
N	Y	Y	N	Y	N
A	B	B	B	D	A
C	D	D	C	A	C
B	C	A	D	C	D
D	A	C	A	B	B

Here the design differs from a fully randomized design in the sense that each plot has a treatment already applied (no treatment or insecticide) before varieties have been selected randomly, rather than allocating to each plot a treatment and variety combination randomly.

Here, the fields constitute blocks. These are then split into two *whole plots*. The insecticide factor is called the *main* factor. Each whole plot is then divided into four equal-area plots and a variety of potato is allocated randomly. This is called a split plot (or a subplot). Here variety of potato is the subplot factor. In this design the whole-plot effect is completely confounded with the treatment factor, whereas the subplot factor is not confounded (varieties are not

---

Example 3.17
If the results in the above example, after arrangement, were as follows, analyse the results (in kg):

	Field					
	1		2		3	
Variety	N	Y	N	Y	N	Y
A	35.8	36.2	39.5	39.7	36.1	35.1
B	42.4	44.1	38.9	42.7	39.8	34.9
C	37.1	41.5	40.2	38.5	38.1	34.2
D	36.8	36.4	35.1	36.3	38.0	32.5

---

confounded), it is sensible to use the factor we are most interested in as the split plot factor.

Solution 3.17
The analysis involves estimating two different types of variability, that in the whole plot and that in the split plot. The full analysis of variance table is as follows:

	Source	d.f.	ss	ms	F	Prob > F
Whole plot	Fields (blocks)	2	39.99	19.99	1.17	0.4161
	Insecticide	1	1.35	1.35	0.08	0.9015
	Field × insecticide (whole plot residual)	2	34.09	17.04		
Split plot	Variety	3	69.71	23.24	5.84	0.0107
	Insecticide × variety	3	2.66	0.88	0.30	0.8789
	Split plot residual	12	47.73	3.98		
	Total	23				

Here, in the top part of the table the estimate of whole-plot variance is used in the F-tests, whereas in the lower part of the table the estimate of split-plot variance is used.

The resulting analysis shows that the only statistically significant effect is between the varieties of potato.

3.5   MULTIPLE COMPARISONS

One of the problems mentioned earlier was that of deciding, once we had ascertained a significant difference existed, which treatment levels differed from which within a group of treatments. A number of different approaches can be used, the main ones being outlined below.

3.5.1   LEAST SIGNIFICANT DIFFERENCE

This is the simplest approach for comparing means of treatments. However, it is also the least reliable since we are carrying out multiple comparisons simultaneously, each with the same significance level α, typically 5%. This means the risk of making an error in the comparison is much larger overall than simply this significance level. A further problem that sometimes arises is that having found a significant result overall in the F-test, no individual differences can be identified.

To find which means are different from which, calculate the least significant difference (LSD) as

$$\text{LSD} = t_\nu s \sqrt{(1/n_i + 1/n_j)} \qquad (3.3)$$

where s is the square root of the residual mean square in the ANOVA table and where t is a t-distribution value with residual degrees of freedom ν and tail probability equal to half of the significance level of the test required, i.e. a two-tail test. n_i is the number of replicates of treatment i, n_j is the number of replicates of treatment j. Any two means separated by this LSD or more are then identified as being different. Usually, n_i and n_j would be planned to be the same although this is not necessary.

In the remainder of this section it will be assumed n_i is the same for all treatments although formulae do exist for cases where the n_i are not equal. (Please refer to the references at the end of the chapter if such formulae are required.)

Example 3.18
In Example 3.5 it was identified that some of the four contaminants were different to others. Determine any differences between individual contaminants using least significant differences.

Solution 3.18
The least significant difference at the 5% level is given by

$$\text{LSD} = t_{12,\,0.025}\, s\sqrt{(1/4 + 1/4)}$$
$$= (2.179) \sqrt{(4.4375)} \sqrt{(0.5)}$$
$$= 3.246$$

The means for the contaminants in decreasing order are

	A	C	D	B
Mean	52.50	49.75	49.25	44.75

so we can say that A and D, A and B, C and B, D and B are significantly different since these means differ by more than the LSD value (3.246).

3.5.2 DUNCAN'S MULTIPLE RANGE TEST

Duncan's multiple range test is an adaptation of the LSD approach where allowance is made to ensure the significance level of the tests overall is at least α.

In this test a variation of the LSD approach is used where the treatment means are first put in ascending order. Then, Duncan's table of significant ranges need to be used. (These are quoted in most books of statistical tables). If

$r_{m, v, \alpha}$ is used to represent these values where m is the number of treatments being compared, v is the residual degrees of freedom and α the required significance level, then a sequential set of comparisons are made. First, compare the largest mean with the smallest mean. The least significant range is given by

$$R_m = r_{m, v, \alpha} \sqrt{(s^2/n)} \qquad (3.4)$$

where n observations of each treatment have been made. Next, compare the largest mean with the second smallest in the same way

$$R_{m-1} = r_{m-1, v, \alpha} \sqrt{(s^2/n)} \qquad (3.5)$$

and so on until the final comparison between the largest and second largest mean with

$$R_1 = r_{1, v, \alpha} \sqrt{(s^2/n)} \qquad (3.6)$$

Wherever the difference in means is greater than or equal to this least significant range, the means are significantly different. This process is then repeated using the second largest mean and so on until all comparisons have been made.

Example 3.19
For Example 3.5 use Duncan's multiple range test to investigate which contaminants lead to differing yields.

Solution 3.19
The means in descending order are:

	A	C	D	B
Mean	52.50	49.75	49.25	44.75

To compare largest to smallest (A with B) use

$r_{4, 12, 0.05} = 3.313$

which gives

$R = (3.313) \sqrt{(4.4375/4)}$
$\quad = 3.489$

The difference between largest and smallest mean is greater than this range so there is difference between the means for contaminants A and B.
   Continuing for the other pairs in order:

A and D, C and B $\quad r_{3, 12, 0.05} = 3.225, R = (3.225) \sqrt{(4.4375/4)}$
$\qquad\qquad\qquad\qquad\qquad\qquad = 3.397$

The means of contaminants A and D do not differ by more than R, those for C and B do.

A and C, C and D, D and B $r_{2, 12, 0.05} = 3.082, R = (3.082) \sqrt{(4.4375/4)}$
$$= 3.246$$
Hence, only means of contaminant D and B are separated by more than R.

Overall, we have a different result to LSD, with B producing a lower yield than the other three contaminants.

This procedure is an effective approach for detecting differences where they really exist and is frequently used.

3.5.3   TUKEY'S TEST

Tukey's test procedure lies between the two previous approaches. It is based on a Studentized range statistic, but there is only one range statistic used for all the paired comparisons. Two means are significantly different if their difference is greater than or equal to

$$Q = q_\alpha (m, v) \sqrt{(s^2/n)} \tag{3.7}$$

where values of q are obtained from Studentized range tables.

This test is more conservative than Duncan's multiple range test but is easier to apply. It will not identify real differences as readily as Duncan's approach.

Example 3.20
For Example 3.5 use Tukey's test to investigate which contaminants lead to differences in yield.

Solution 3.20
Here the table value is

$q_{0.05} (4, 12) = 4.20$

giving Tukey's Studentized range of

$Q = (4.20) \sqrt{(4.4375/4)}$
   $= 4.424$

We can see here that this range leads to an ordering between the four means of

A, C, D > B

since the range used for comparison is rather larger. Usually more conservative than Duncan's, this example produces an identical result.

3.5.4 DUNNETT'S TEST

When we have a control that is to be compared to a series of treatments the number of treatment-control comparisons is $m - 1$. To make these comparisons a modification of the earlier least significant difference approach is used where the t-value is replaced by a value from Dunnett's tables. Here, the joint significance level associated with all $m - 1$ tests is α, rather than for each individual test.

The least significant range is given by

$$D = d_\alpha (m - 1, v)\sqrt{(2s^2/n)} \qquad (3.8)$$

where d is the value from Dunnett's table.

Example 3.21
For Example 3.5 use Dunnett's test to investigate which contaminants lead to differences in yield from a control if contaminant B is, in fact, a control.

Solution 3.21
The value from tables is

$d_{0.05} (3, 12) = 2.68$

and the range used for comparison is

$D = (2.68) \sqrt{(2 \times 4.4375/4)}$
$\quad = 3.992$

Here we see that all the three means for contaminants A, C, D are more than this range from the mean for B (control) so all contaminant means are greater than the control mean.

This procedure is specifically designed for comparing one individual treatment (usually a control) against all other treatments.

3.6 CONTRASTS

Contrasts are a means of testing theories related to individual treatments or groups of treatments. If we have six treatments A, B, C, D, E, F we may for instance wish to investigate whether A and B are the same. We may wish to test whether A and B are the same as D and F. The contrasts of interest would have been identified in advance of obtaining the observed results. By adapting the ANOVA table such contrasts can be tested.

3.6.1 ORTHOGONAL CONTRASTS

A contrast is defined as

$$C = \sum_{i=1}^{m} c_i y_{i\bullet}$$ (3.9)

where c_i is a contrast weight and $y_{i\bullet}$ the ith treatment mean. For this to be a valid contrast this linear combination of treatment means must have weights which add up to zero, i.e.

$$\sum_{i=1}^{m} c_i = 0$$ (3.10)

In the two examples above we could use the following weights:

TREATMENT WEIGHTS

Contrast	A	B	C	D	E	F
A vs B	1	−1	0	0	0	0
A, B vs D, F	1	1	0	−1	0	−1

(notice the signs within a contrast could be reversed).
A sum of squares for each contrast of interest is calculated as

$$SS_c = \frac{\left(\sum_{i=1}^{m} c_i y_{i\bullet}\right)^2}{n \sum_{i=1}^{m} c_i^2}$$ (3.11)

This has 1 d.f. and is tested against the residual mean square as usual. In effect, the treatment sum of squares is being decomposed into parts due to particular contrasts. If we wish to investigate more than one contrast simultaneously, then we would want the contrast tests to be independent of one another. This is achieved by choosing contrasts which are orthogonal to one another.

If the weights in one contrast are c_i and in another d_i, the two contrasts are orthogonal if

$$\sum_{i=1}^{m} c_i d_i = 0$$ (3.12)

The number of orthogonal contrasts which can be compared simultaneously equals the number of degrees of freedom for the treatments. For the six treatments above there are five contrasts we could investigate. Identifying which set of contrasts are all orthogonal can be difficult and often requires trial and error. The two contrasts already given are orthogonal and a possible set of further contrasts would be

TREATMENT WEIGHTS

Contrast	A	B	C	D	E	F
A vs B	1	−1	0	0	0	0
A, B vs D, F	1	1	0	−1	0	−1
A, B vs C, E	1	1	−1	0	−1	0
D vs F	0	0	0	1	0	−1
C vs E	0	0	1	0	−1	0

Each pair forms an orthogonal contrast so all these contrasts can be compared simultaneously.

Example 3.22

For Example 3.5 carry out an orthogonal contrast analysis if the investigator wishes to compare contaminants A and B against contaminants C and D, the reason being that A and B are natural contaminants, whereas contaminants C and D are synthetic.

Solution 3.22

First we need to construct the contrasts.

Constrast	A	B	C	D
A, B vs C, D	1	1	−1	−1
A vs B	1	−1	0	0
C vs D	0	0	1	−1

And the contaminants' sums of squares can be decomposed as follows:

Source	d.f.	ss	ms	F	Prob > F
A, B vs C, D	1	3.06	3.06	0.69	0.4223
A vs B	1	120.12	120.12	27.07	0.0002
C vs D	1	0.50	0.50	0.11	0.7429
Contam. total	*3*	*123.69*	*41.23*	*9.29*	
Residual	12	53.25	4.44		
Total	15	176.94			

This suggests that, of the three contrasts, only the A vs B comparison shows a difference between these contaminants.

3.6.2 SCHEFFE'S TEST

Scheffe's test is for comparing contrasts when the experimenter cannot or does not identify contrasts of interest in advance. The overall significance level of any comparison is at most α. Calculation of Scheffe's test is a complex process based on contrasts and best left to computer procedures such as those within the computer package. In many practical situations we will only be interested in comparisons of pairs of treatments. Using the example 3.5 leads to the suggestion that contaminants A and C are both significantly higher yielding than B. This approach then becomes an alternative to the tests used in section 3.5 (see Sokal and Rohlf 1995).

3.6.3 POLYNOMIAL CONTRASTS

In many ecotoxicological situations the treatments form a sequence of concentrations from zero (the control) upwards. The specific concentration levels may in themselves not be important, rather it is the identification of a response, and the shape of that response that is of particular interest. Where there are k levels of a contaminant, including the control as a zero concentration, we can

---

Example 3.23
Perennial ryegrass plants have been exposed to three concentrations of lead: 15, 30 and 45 ppm. There are three replicate plots of each dose. The length of the longest root of four plants in each plot has been measured and averaged. Data are as follows:

Dose	Length (mm)
15 ppm	12.3 10.3 8.3
30 ppm	11.0 4.8 9.3
45 ppm	2.5 1.3 1.0

Undertake an analysis of variance of these data and examine the shape of the response (if any) using orthogonal polynomial contrasts.

Solution 3.23
The usual ANOVA of these data would be as follows:

Source	d.f.	ss	ms	F	Prob > F
Treatment	2	125.22	62.61	12.61	0.0071
Residual	6	29.79	4.96		
Total	8	155.00			

but the treatment effect can be broken down into its linear and quadratic components

Source	d.f	ss	ms	F	Prob > F
Linear	1	113.54	113.54	22.89	0.0030
Quadratic	1	11.68	11.68	2.35	0.1762
TRT total	*2*	*125.22*	*62.61*	*12.61*	
Residual	6	29.79	4.96		
Total	8	155.00			

This clearly demonstrates that the response, over this range of doses, is approximately linear with little evidence of curvature.

decompose the treatment sums of squares with $k - 1$ degrees of freedom into $k - 1$ orthogonal polynomial contrasts. For example, if we had four dose levels we could break the treatment sums of squares into its linear, quadratic and cubic components and determine the shape of the response (if any) to increasing dose.

3.7 FURTHER ISSUES

3.7.1 DEVELOPMENT OF DESIGNS

The experimental designs outlined in the chapter are only a selection of those available to the researcher or investigator. All the designs discussed can be generalized and extended to more factors. Some can be combined with others. Some can be modified. An example is the balanced incomplete block design where treatments are balanced but where not all treatments can be applied in a single block.

Example 3.24
The following layout would be a balanced incomplete block design:

Treatments	Blocks					
	I	II	III	IV	V	VI
1	x	x	x		x	
2	x		x	x		x
3		x		x	x	x

It contains three treatments in six blocks of two plots, leading to two replicates.

A further example might be a fractional factorial design where only some of the factors which form the experiment are to be analysed, others being allowed to be confounded.

A statistician will be able to advise you of the best design to use for your particular investigation, but only if she/he is involved at an early stage!!

3.7.2 SAMPLE SIZE CONSIDERATIONS

An essential part of the experimental design process is a decision on the extent of recording. Chances of detecting effects are improved by (1) reducing the residual variation of a trial (through precision recording, sampling intensity, blocking, covariates, etc.) and (2) increasing the replication. Increasing replication is an expensive process, particularly in field situations. Many eco-toxicological measurements are associated with high cost and high variability, yet it is these situations where increased replication is so often necessary. Post-

Table 3.1 Approximate detection effects (two standard errors of the difference expressed as a percentage of mean) for replications of 2, 4 . . . 10 and CVs of 5, 10, 15 and 20%. The power of these tests to detect differences is not high (about 50%); to gain power more resources have to be put into experimentation. CVs may be higher than these in many typical ecotoxicological situations, further emphasizing the need for adequate replication

		Coefficient of variation (%)			
		5	10	15	20
Replication	2	13	26	39	51
	4	8	15	23	30
	6	6	12	18	24
	8	5	10	15	20
	10	4	9	13	17

mortem examination of results can identify what level of effect could be detected for a given level of precision. This process can be a real eye-opener. For example Table 3.1 shows the effect (expressed as a percentage of the mean) that can be detected as statistically significant for varying replication and residual variation levels. Here the importance of reducing variability and replication in detecting effects likely to be of biological importance can be clearly seen.

Often the level of replication is constrained by experimental limitations. In these situations it may be possible to reduce variability by intensive sampling within each experimental unit. An experimental facility to examine the effects of elevated CO_2 on plants and invertebrates is restricted to two replications (domes) and in each replicate a number of winter moth caterpillars were raised.

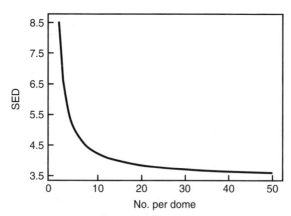

Figure 3.4 In a fixed replicate experiment, the standard error of the difference (SED) can be reduced by increasing the measurement of individuals within the replicate. However there is a lower limit (asymptote to which the SED can fall, further improvement can only occur through increased replication.

Figure 3.4 shows the consequences of varying the number of caterpillars recorded in each dome. Reductions in the SED (standard error of the difference) are obvious as caterpillar numbers increase. There is a point at which little increased precision can be gained, probably at about 10 caterpillars in this example. Please note that increasing caterpillar numbers does not increase experimental replication, just the accuracy with which each experimental unit is measured.

Where replication is limited in this way, additional replication may be achieved by either utilizing other facilities elsewhere, or by undertaking replicates sequentially through time. The latter obviously requires a different randomization for each sequential replicate.

Studying the effects of varying both replication and sampling intensity can be of extreme value. Sadly this practice is not commonplace.

3.7.3 COVARIABLES

We do not have space in this introductory text to consider this important area in detail. A covariable can be considered to be a variable that is influencing the outcome of an experiment. As such it is a nuisance variable that is likely to be increasing the residual variation and reducing the chances of detecting statistically significant effects. It is possible to include a covariable in the designs discussed above and hence eliminate it from our comparison of treatments (effectively we are removing another source of variation). In ecotoxicological examples our plots might consist of worms or animals of different sizes or ages and hence of different susceptibility. If this information is recorded it can be eliminated as a nuisance variable from our analysis.

3.7.4 PITFALLS WHEN USING DESIGNED EXPERIMENTS

If the design is chosen appropriately at the outset of the investigation then most of the problems centre around the nature of the data collected and the assumptions we made earlier (section 3.4) about these data. The assumption of independence of observations is an important one to consider even at the planning stage. Where plots are being used, consideration needs to be given to avoiding a plot treatment accidentally being partially applied to an adjacent plot. The way the data themselves are measured and collected must ensure one result does not influence any other. If independence of results is not maintained it is likely the results and conclusions will be biased.

The assumption of equality of variance regardless of the result itself can often be violated in practice. It is not uncommon, for instance, for the variability of results to increase with the magnitude of results. In these circumstances it is possible to use transformations of the original data before it is analysed. Common transformations used are log, square root and reciprocal of the data.

The final assumption concerned normality of the data. If the data come from a population which is skew then we cannot immediately assume normality. Once again transformations might be used to convert the data to normality. Where normality cannot be achieved then non-parametric test equivalents of analysis of variance need to be used. The two major tests are the Kruskal–Wallis and the Friedman tests.

For further information about statistical experimental design there are some excellent texts devoted to this topic. The following are recommended.

REFERENCES

Montgomery DC (1991) *Design and Analysis of Experiments*, Wiley, New York.
Johnson NL and Leone FC (1990) *Statistics and Experimental Design in Engineering and the Physical Sciences*, Wiley, New York.
Sokal RR and Rohlf FJ (1995) *Biometry*, W. H. Freeman, New York.

4

The Analysis of Laboratory Toxicity Experiments

REINHARD MEISTER[1] AND PAUL J. VAN DEN BRINK[2]

[1]*Technische Fachhochschule Berlin, Germany*
[2]*DLO Winand Staring Centre, The Netherlands*

4.1 INTRODUCTION

In our industrialized world, large amounts of chemicals are used. They can enter the environment via air, water and/or soil and may have undesirable effects on the biological and functional properties of environmental resources and on human health. In order to prevent unacceptable consequences of the use of chemicals, the possible extent of their effects on the environment and on humans should be estimated before they are released into the environment. To ascertain, for example, whether a chemical has a carcinogenic effect on humans, tests with mammals (usually rats) are performed. In this case the rat serves as a model for man.

To evaluate possible hazards on exposed ecosystems, one can conduct field tests as described in Chapter 3. But if one considers that in The Netherlands alone approximately 250 different pesticides are permitted, testing every allowed chemical with the help of expensive and elaborate field experiments can be considered impossible.

It would therefore be helpful to be able to ascertain the toxicity of a chemical using simple tests, in a manner comparable to mammalian toxicology. The OECD (Organization for Economic Co-operation and Development 1993) has developed guidelines for the testing of chemicals. Their expert group on ecotoxicology developed tests for the following species:

Acquatic ecosystems
    Algae, growth inhibition test
    Daphnia sp. acute immobilization test and reproduction test
    Fish, acute toxicity test, prolonged toxicity test, early-life stage test

Statistics in Ecotoxicology. Edited by T. Sparks. © 2000 John Wiley & Sons Ltd

Terrestrial ecosystems
     Terrestrial plants, growth test
     Earthworm, acute toxicity tests
     Avian dietary toxicity test, reproduction test
Treatment plants
     Activated sludge, respiration inhibition test

For the sake of hazard assessment of chemicals in aquatic ecosystems, a simplified ecosystem is considered which contains only three species: an algae, a water flea and a fish. In this 'hazard assessment ecosystem' the algae represents the plants, the water flea the herbivores and the fish the predators. Most tests distinguish between endpoints, e.g. inhibition of growth, survival, immobility or inhibition of reproduction may be considered.

The simplification of the aquatic ecosystem also has great disadvantages. By reducing the cost in terms of logistics, space, time and money, one also decreases the ability to extrapolate the results to a field situation. By performing laboratory tests one totally ignores the interactions between species, between species and their environment and between the stressor and the environment. On the other hand, laboratory results are very easy to interpret and very accurate because of the controlled environment in which the test is conducted.

Until now we have focused on the use of laboratory tests for hazard assessment, regulatory purposes. Laboratory tests are also in use for more scientific goals. Among other things, one often needs the results of laboratory tests to be able to interpret the results of field tests. A simple question like 'Is the decrease in numbers of a certain species as observed in a field test a direct result of the stressor or an indirect effect due to species interactions?' can only be answered with the help of toxicity data. Sometimes the processes to be investigated are of such complexity that one has to start with one organism in the laboratory. A good example is the research on the joint toxicity of mixtures of chemicals. Laboratory tests can also be of use when investigating, for instance, the bioconcentration or biokinetics of a certain chemical in a species. For simplicity we will mainly focus on aquatic laboratory tests in this chapter. The results of such tests can be used to develop standards for the 'safe concentration' of a pesticide in a water body adjacent to an agricultural field.

4.2   DESIGN OF LABORATORY TOXICITY TESTS

As noted above, for cost-effectiveness and practicability the aquatic ecosystem was reduced to three species. In order to ascertain their susceptibility to a single compound or complex mixtures like waste waters, several individuals of a species are exposed in a test vessel to the test compound or the mixture. If one is interested in mortality, the mortality occurring in this test vessel is compared

Table 4.1 Toxicity of Linalylacetate on *Daphnia magna*.
(Data from Dr Steinhäuser, Umweltbundesamt Berlin)

Concentration	Died within 48 h	n at start
0	0	20
9	0	20
11	0	20
13	0	20
16	0	20
19	0	20
23	0	20
28	0	20
33	3	20
40	5	20
48	9	20
58	12	20
69	13	20
83	19	20
100	20	20

with those occurring in a control test vessel to which no chemical is added. Because one does not know a priori at which concentration certain effects occur, several concentrations are tested simultaneously in this way. By noting the mortality occurring at all concentrations, a relation between the dose and the magnitude of the effect (in this case mortality) is recorded. In acute toxicity tests, two important parameters can be calculated from the obtained dose–response relation: the EC_{50} and the NOEC.

LC_{50}/EC_{50} (lethal concentration 50%, effect concentration 50%): The concentration of a test substance which produces the investigated response in 50% of the test organisms.

NOEC (no observed effect concentration): The highest concentration of a test substance in a toxicity test that has no statistically significant adverse effect on the population of exposed test organisms as compared to the control (Rand and Petrocelli 1985).

Table 4.1 gives the data of an experiment where the survival of water flea, *Daphnia magna*, was studied when exposed to different concentrations of the organic solvent, Linalylacetate. In this experiment for each concentration two replications each of 10 daphnids were used. The number of survivors after 24 and 48 h were recorded. In Table 4.1 we give the combined data for the total two-day period. Based upon visual inspection of the data, the LC_{50} lies between 48 and 58 $\mu g \, l^{-1}$, and the NOEC will be 28 or 33 $\mu g \, l^{-1}$. From the literature it is known that the results of these tests, even when performed with the same species and test substance, may differ considerably (by up to an order of magnitude; Rand and Petrocelli 1985) between laboratories. A few possible, obvious, causes of these differences are briefly mentioned below:

Differences between populations tested

The populations tested may differ with regard to their demographic structure. Younger and therefore smaller individuals seem to be more sensitive compared to adult/larger ones (Rand and Petrocelli 1985). This is a result of the larger surface/capacity ratio of smaller individuals, because a larger amount of the test chemical is absorbed per amount of body mass. Moreover, young individuals have a relatively higher respiration rate compared to adult individuals, because of a higher metabolic activity per unit of body weight.

The history of a population can also be of importance. The population tested can, for instance, be exposed to a stressor before the execution of the test. This stressor may have killed the weaklings in the population at a younger age. The consequence is that the remaining older, sturdier animals remain. Furthermore, geographic, genetic variation between populations will result in differences in susceptibility.

Differences in test conditions

Surrounding environmental factors may also influence the test results. Temperature, for instance, may influence the general activity, level of metabolism and behaviour of aquatic organisms. It may also alter the physico-chemical characteristics of the chemicals, for instance its half-life, and therefore its behaviour.

The quality of the accommodation of the test organisms, and the experimental researcher can also be important. Small test vessels or the absence of typical habitat requirements may lead to physical stress, which may also alter the chemical stress. Differences in accuracy of the operator can also lead to a different outcome of the test.

Differences in experimental set-up

A test with a duration of one day generally results in a higher LC_{50} compared to a four-day test. Some time is needed to establish equilibrium between a compound concentration at its site of action and its concentration at the site of application, i.e. in the surrounding water. Thus some time is needed for a LC_{50} value to reach its incipient value; the asymptote value of the curve of toxicity against time. The number of concentrations, replicates and organisms per test vessel, and the precision of test, may of course also alter its results.

In order to minimize the differences in results the OECD developed some guidelines (as mentioned above) to standardize protocols for the cultivation of

Table 4.2 Summary of properties of standard laboratory tests (strongly simplified) as recommended by the OECD (OECD 1993)

	Algae	Daphnia sp.	Fish (excluding early-life stage)
Duration	3 days	1–4 days (acute test) 14 days (reproduction test)	4 days (acute test) 14 days (prolonged test)
Initial density	10^4 cells ml^{-1}	20 ind/test unit	1 g fish l^{-1}
Test species	Selenastrum capricornutum Scenedesmus subspicatum Chlorella vulgaris	Daphnia magna (#/24 h) Daphnia sp. (#/24 h)	Several, e.g. Poecilia reticulata (guppy) Ind. of 2 ± 1 cm.
Temperature	21–25 °C	18–22 °C	Depends on species, for guppy: 21–25 °C
Light regime	120: E/m^{-2} s^{-1}	Not necessary (acute) 16 h photoperiod (reproduction)	12 to 16 h photoperiod
Endpoint	Growth rate	Survival (acute) Immobilization (acute) Reproduction (reproduction)	Survival
Results	EC$_{50}$ NOEC	LC$_{50}$ (acute) EC$_{50}$ (immobilization, acute) NOEC (reproduction)	LC$_{50}$ (acute) NOEC (prolonged)

test organisms, the conditions of the experiment, the experimental set-up and the data analysis (see Table 4.2 for a summary).

4.3 DATA ANALYSIS

4.3.1 INTRODUCTION

As mentioned above, data from laboratory toxicity experiments all have basically the same structure. Several suitably chosen concentrations or doses including a zero dose as a control define the experimental groups. The experimental units are allocated at random to the different groups. This process is called randomization, and is an essential for the reliability of all types of inference drawn from the results of the experiment.

Figure 4.1 shows a plot of the data given in Table 4.1 and it is obvious that the ratio of dead to number of *Daphnia* at start increases with concentration.

These kinds of data, where only two types of reactions can occur for each individual, are called quantal-response data. When investigating quantal-response data one is usually interested in the probability of an event (in this example death of a daphnid within a two-day period) and the relation of this probability to the exposure level. There will be a formal introduction in quantal response models in section 4.3.2.

The next example deals with quantitative response data. In this example the

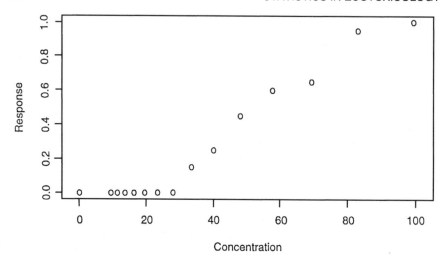

Figure 4.1 Ratio of dead/(number of *Daphnia* at start) for different concentrations of Linalylacetate.

relative change in volume of algae, rather than the probability of death, was taken as the endpoint. Two replications per concentration were analysed and the relative change in volume on day 3 is reported. For these data, a sharp decrease in volume is observed with increasing concentration. The original values are given in Table 4.3. We will look for a suitable model which can be fitted to this type of data in section 4.3.3

The following two sections will discuss some suitable models to describe data similar to our introductory examples. The presentation is kept at a very basic mathematical level. The main purpose is to give the reader a feeling of what is possible when analysing data from ecotoxicological experiments. As quantal-response models are usually not covered in introductory statistics, this section gets a little more space.

Table 4.3 Toxicity of a herbicide for the algal species *Scenedesmus acutus*. (Data from Paul J. Van den Brink)

Concentration	Relative volume on day 3	
0.00	100.91	111.22
0.89	101.48	90.78
8.94	43.72	35.84
40.00	3.81	7.01
78.61	3.81	3.61
98.11	0.91	1.54
196.22	0.65	0.19

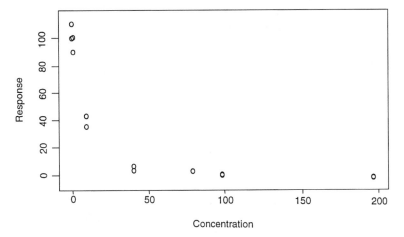

Figure 4.2   Relative volume on day 3 of the algal species *Scenedesmus acutus* exposed to different concentrations of a herbicide. Data from Paul J. Van den Brink.

4.3.2   THE ANALYSIS OF QUANTAL-RESONSE DATA

4.3.2.1   Quantal response in a hypothetical setting

We return to our first example where the proportion of dead water fleas increased with concentration of a substance. How can a theoretically based model explain the experimental data? We will consider a similar problem, where the solution is completely known. This is a kind of mental exercise which helps us to understand the approach. We assume that we have N water fleas which are able to resist individually varying concentrations of a pesticide (c), defined as their tolerance T. Once the concentration exceeds the individual resistance, the animal will die. Our hypothetical data are shown in Table 4.4.

We generated Table 4.4 using pseudo-random numbers from a standard normal distribution. For the individual tolerances, T, a log-normal distribution was assumed. To keep the numbers as simple as possible, the logarithm to base 2, i.e. $\log_2$, was used instead of either the natural log, $\log_e$ or $\log_{10}$ which are commonly used in practice. The values of T were generated by the equation $\log_2(T) = 2 + Z$, where Z was taken from a normal $N(0, 1)$ distribution.

If we know the individual tolerances, there is no need to perform a concentration response experiment because all animals with tolerances below the experimental concentration would die. The best description of our data can be given in terms of mean and standard deviation of the transformed values $Y = \log_2(T)$. For the example we obtain mean(Y) = 2.08 and SD(Y) = 0.93 in agreement with the mechanism of generation. (The theoretical mean and standard deviation of $Y = \log_2(T)$ are 2 and 1 respectively.)

In real experiments we do not know what the individual tolerances are, the

Table 4.4 Individual tolerances, T, of $N = 50$ animals assigned at random to one of five groups, $x = 1, 2 \ldots 16$. Hypothetical data, values below the experimental concentration in bold

$c = 1$	$c = 2$	$c = 4$	$c = 8$	$c = 16$
1.43	**0.82**	**2.08**	**1.61**	**1.23**
1.96	**1.45**	**2.81**	**1.73**	**3.79**
2.75	2.05	**3.44**	**3.31**	**3.99**
3.33	2.23	**3.61**	**4.49**	**4.23**
4.57	2.27	**3.63**	**4.81**	**4.45**
4.66	4.15	5.38	**4.96**	**4.84**
4.80	5.21	5.47	**6.81**	**5.34**
6.83	5.23	6.46	**7.45**	**5.86**
11.07	8.18	8.36	9.81	**6.21**
13.12	11.13	8.89	17.17	**7.13**

only information available is whether or not an animal dies. We introduce a new variable X which takes the values 0 and 1, where $X = 1$ denotes the death of an animal. We have $X = 1$ when $T \leqslant c$, an animal dies when exposed to a concentration larger than its tolerance and, therefore, the probability for death given the concentration c is

$$\Pr(X = 1 \mid c) = \Pr(T \leqslant c) \tag{4.1}$$

As our tolerances are $\log_2$-normally distributed we get, using the $\log_2$ transformation,

$$\begin{aligned}
\Pr(T \leqslant c) &= \Pr[\log_2(T) \leqslant \log_2(c)] \\
&= \Phi[(\log_2(c) - \mu)/\sigma] \\
&= \Phi[(\log_2(c) - 2)/1]
\end{aligned} \tag{4.2}$$

where Φ denotes the cumulative distribution function of the standard normal distribution, taking the theoretical values 2 and 1 for μ and σ. This model is often called the Probit model, Bliss (1935) introduced this term as 'probability unit'. Today the inverse of the normal-distribution function Φ, is called the Probit-function.

We summarize our hypothetical data from Table 4.4 according to a quantal-response study and obtain the data given in Table 4.5.

How can we estimate the parameters of the Probit model using the data from Table 4.5? This is the task of quantal-response models. Estimation is usually performed using maximum-likelihood (ML), a method which chooses parameter estimates that maximize the probability of the observed responses. The likelihood is easily determined as the number of reactions (deaths) per concentration is binomially distributed and the response probability depends on unknown parameters μ and σ according to equation (4.2). Usually the task of ML estimation is taken by specialized statistical software.

Table 4.5 Quantal response summary of data from Table 4.4, where c denotes concentration, n the number of exposed, r the number of dead and the last column gives r/n as a percentage

c	n	r	%
1	10	0	0
2	10	2	20
4	10	5	50
8	10	8	80
16	10	10	100

A closer look at the data from Table 4.4 shows that the concentration–response relationship has an apparent point of symmetry at $c = 4$. In addition the $\log_2$ transform of c seems natural. So we may choose $\log_2(4) = 2$ as an estimate for μ. This is in perfect agreement with the theoretical value. However, in practice, there is little hope that an experimental concentration equals the true centre of symmetry. Now we try to find an estimate for σ. If we assume that equation (4.1) is correct, take $\mu = 2$, and replace the theoretical probabilities by the observed proportions r/n we find for $c = 8$, where $\log_2(8) = 3$,

$$8/10 = \Phi[(3 - 2)/\sigma] \tag{4.3}$$

Now we use the inverse of Φ, called Probit and obtain

$$\text{Probit}(0.8) = 1/\sigma \iff \sigma = 1/\text{Probit}(0.8) \tag{4.4}$$

and end up with $1/0.84 = 1.19$ as an estimate for σ, not too far from what we expected. Values for the Probit function can be found in all tables giving normal quantiles (e.g. $\text{Probit}(0.975) = 1.96$).

One thing is very important in our rough estimation: in contrast to the calculation of mean and standard deviation, the estimation of σ from quantal-response data was only possible using a model for the expected responses. Using a different model can yield different estimates.

The results from the ML estimation using the Probit model for these data are $\mu = 2$ and $\sigma = 1.04$ respectively.

4.3.2.2 Theory and practice of quantal-response analysis

This section gives a brief introduction of the statistical techniques necessary to analyse quantal response data.

What do we expect from a concentration response experiment? In ecotoxicology, we want to relate the observed response with the concentration of the toxic exposure. More precisely stated, for quantal response we want to know what is the probability of response for a given concentration or, vice versa, at which concentration do we expect a given response probability. Once we have this information, a risk assessment can be performed comparing expected environmental concentrations and predicted response probabilities.

4.3.2.2.1   Quantal-response models

We consider models of the type

$$\Pr(X = 1 \mid c) = p(c) = F(\{\log(c) - \mu\}/\sigma) \qquad (4.5)$$

where the probability of a reaction is determined by a distribution function F, a location parameter μ and a scale parameter σ. Once the distribution type F is chosen and μ and σ are known it is easy to calculate the response probability for a given concentration. One has just to plug in the known values into the model equation. This simplifies the calculation of so-called effective concentrations. The EC_{100p} values give a concentration such that

$$p(EC_{100p}) = p \qquad (4.6)$$

holds. For symmetrical functions, F, with $F(0) = 1/2$ we have always

$$\log(EC_{50}) = \mu \qquad (4.7)$$

which means that the location parameters equals the log of the 50% effective concentration. In general we get, after some basic algebra,

$$\log(EC_{100p}) = \mu + \sigma \times x_p, \qquad (4.8)$$

where x_p denotes the p-quantile of the distribution function F with $F(x_p) = p$. Effective concentrations for small response probabilities, for example an EC_5 or EC_{10}, value are sometimes proposed as input for a risk assessment strategy.

Different model types are used in practice, but in many applications the results do not differ very much between these models. There is the probit model, where the function F is the standard normal distribution function. Using the logistic distribution for F gives the so-called logistic-regression model. Another type is the Weibull model. Figure 4.3 demonstrates the role of the scale parameter σ in such models.

It is obvious that small values of σ, indicating a small biological variability, give a very sharp increase of the concentration–response curve. As the logistic function is symmetric, we see that the location parameter, μ, corresponds to $\log(EC_{50})$. In Figure 4.3 the EC_{50} value was kept constant and equal to 1 for all values of σ.

4.3.2.2.2   Maximum likelihood estimation for quantal-response models

In section 4.3.2.1 we have already demonstrated how the unknown parameters can be estimated from the data. Now we will discuss this topic in more detail.

When fitting models to data, one has to think about the quality of fit. In other words one has to define what a good fitting model is. From ordinary regression,

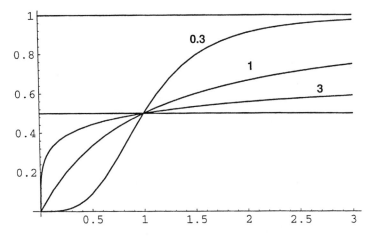

Figure 4.3 The log-logistic model for different values of σ. Vertical axis $p(c)$, horizontal axis c.

we know that the least squares criterion gives such a definition of good fit. However, the least squares criterion cannot always be applied. For quantal response data the ML criterion is usually applied. What does this mean?

The idea behind ML estimation is very simple: The probability of the observed data is taken as the criterion for a good fit. This probability is called likelihood. Given the data and a model, the likelihood depends only on the model parameters. The best fit is achieved if the parameters are chosen such that the likelihood takes its maximum. These parameter values are called ML estimates. We do not give the formula for the likelihood here. Interested readers should consult one of the books on quantal-response methods, e.g. Morgan (1992).

There are several theorems in mathematical statistics which support the use of ML estimates. Other methods developed in the pre-computer age are, in general, less efficient and therefore not presented here. These theorems state, that asymptotically (for large samples) using ML estimates makes the most of our data. In theory there is no superior method for analysis. From the estimation one also obtains all the information necessary to calculate goodness of fit criteria and confidence intervals.

4.3.2.3 Analysis of a quantal-response data set

We will use the data presented in Table 4.1 to answer the questions on the acute toxicity of Linalylacetate for *Daphnia*. Using the Weibull model which has the form

$$F(c) = 1 - \exp\left(-\exp[\{\log(c) - \mu\}/\sigma]\right) \qquad (4.9)$$

or equivalently

$$F(c) = 1 - \exp(-\{c/\alpha\}^\beta) \qquad (4.10)$$

we get the ML estimates $\mu = 4.13$ and $\sigma = 0.278$. Now we calculate $EC_{50} = 55.9$, and $EC_{10} = 33.1$. To do this we use

$$\log(EC_{100p}) = \mu + \sigma \times \log(-\log(1 - p)) \qquad (4.11)$$

which corresponds to the general formula given in section 4.3.2.2.1. To verify that $x_p = \log[-\log(1 - p)]$ is really the desired p-quantile of the Weibull model we plug this expression into the model equation and get

$$F(x_p) = 1 - \exp[-\exp\{\log(-\log\{1 - p\})\}] = p \qquad (4.12)$$

We get for example

$$\log(EC_{10}) = 4.13 + 0.278 \times \log[-\log(1 - 0.1)] = 3.5 \qquad (4.13)$$

and derive finally

$$EC_{10} = \exp(3.5) = 33.1 \qquad (4.14)$$

As the Weibull model is not symmetric $\exp(\mu)$ does not give the EC_{50}. We obtain the value $EC_{50} = 55.9$ in the same manner as EC_{10}.

Checking whether the fit is sufficient, we compute a χ^2-value derived from the likelihood-ratio statistic and get a $\chi^2 = 8.8$ with 12 d.f. The p-value for this statistic is $p = 0.7183$ indicating that our model is in good agreement with the experimental data. The results of the analysis are presented in Figure 4.4.

Summarizing this section we state: quantal-response data can be analysed using models like the Probit, the logit or the Weibull. For estimation of parameters, the ML method is recommended. A substance has to be regarded as toxic if the slope parameter, which is equal to $1/\sigma$, is significantly different from zero. For the characterization of the toxic potency the use of EC_{100p} values (along with confidence limits) is recommended. In addition to the EC_{50} value, which represents only one point of the concentration response curve, information about concentrations causing smaller effects like EC_{10} are useful to describe the toxic potency of a substance. The computation can be done using standard statistical software, for instance GLIM, SAS or GENSTAT.

4.3.3 THE ANALYSIS OF QUANTITATIVE-RESPONSE DATA

Studying the toxicity of substances is not restricted to quantal endpoints. Many variables of interest are quantitative. It is therefore necessary to provide an approach that can handle toxic effects expressed by quantities such as length, weight, fertility, etc. (see example data from Table 4.2).

For our presentation, we will assume that the concentration-dependent toxic effect causes a decrease in the mean value of the response variable. If

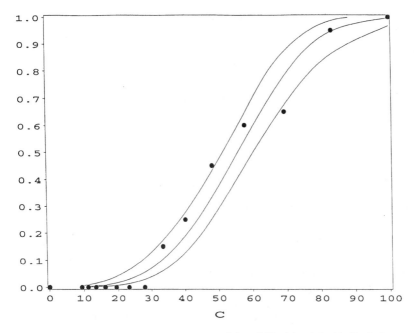

Figure 4.4 Weibull fit to data on the acute toxicity of Linalylacetate to *Daphnia magna*. Observed rates, fitted model with 95% confidence bounds.

we decompose the experimental data into a signal and a noise component, we get

$$Y(c) = \mu(c) + \text{error} \qquad (4.15)$$

where $\mu(c) = E[Y(c)]$ denotes the expected mean of response given concentration c, the signal is also called the systematic component. The error term has expectation 0.

A typical concentration-response experiment has a fixed number of concentrations $0 = c_0 < c_1 .. < c_k$. We have n_i, $i = 0, \ldots, k$ observations $y_{ij}, j = 1, \ldots, n_i$ individuals per concentration group, and the data can be considered being independent if the experimental units (*Daphnia*, fish, . . .) are allocated at random to the groups. Sometimes more complicated structures occur, where observations are blocked, for instance fish in the same tank that might react dependently.

We will explain how changes in $\mu(c)$ can be described using simple regression models. But before this we will discuss briefly the more traditional no-effect approach.

4.3.3.1 The NOEC approach

The idea behind the NOEC (no observed effect concentration) approach is very simple and fascinating. If one believes that there is a smallest concentration below which no change in the mean response is induced by a substance, then there is no risk for the species investigated if an environmental concentration is below this value. Although none of the concentrations tested will equal the true no effect value, one defines the largest of the c_i values for which no statistically significant difference to the control can be established as NOEC. Any test suitable for the comparison of two groups can be used to derive NOEC values. A correction for multiple testing should be made. However, we have to remind the reader that a non-significant result of a statistical test is not at all a proof of no difference in reality.

4.3.3.2 Regression models

Instead of comparing the experimental groups to the control separately, regression models use all available data in order to discover a quantitative relation between concentration and measured response. Often there is no theory at hand that suggests a specific form for the relationship. In this section we explain some simple functions which are often used for the analysis of concentration–response curves. These functions already cover a considerable range of relationships observed in routine testing for ecotoxicity.

We start with the straight line and the parabola, in order to include the simplest possible functions in our presentation. For slight to moderate changes in the observed means, these simple functions can give an adequate fit to the data (see section 4.3.3.3.1).

The Straight Line

The easiest way to describe the change of μ with c is the simple regression equation:

$$\mu(c) = \alpha + \beta \times c \qquad (4.16)$$

It may be regarded as a first-order approximation to the 'true' $\mu(c)$.

Polynomials (the Parabola):

If the function $\mu(c)$ has no constant slope, a polynomial such as the parabola

$$\mu(c) = \alpha + \beta_1 \times c + \beta_2 \times c^2 \qquad (4.17)$$

may give a better approximation to the true curve. Higher-order polynomials should be avoided, usually they are not biologically meaningful outside of the fitted range.

The Logistic Function:

If we expect the mean response to decrease monotonically from a maximum value to an asymptopic value of zero, a line or a parabola are unsuitable models.

Often such functions have a sigmoidal shape. One prominent model with these properties is the log-logistic:

$$\mu(c) = \mu_0/[1 + \exp\{\beta \times [\log(c) - \log(EC_{50})]\}] \qquad (4.18)$$

where μ_0 denotes the maximum achieved for $c = 0$, β determines the slope and EC_{50} is such that $\mu(EC_{50}) = \mu_0/2$ holds. We have already seen the logistic function in connection with quantal response data, where the maximum probability μ_0 equals 1 naturally.

Other Models:

There are many more relationships of the mean response to the concentration of a toxicant than those described above. If the whole range of possible reaction is covered the need for specific models will be large. If the effects are studied only in a limited range, simple descriptive models may be sufficient. The range of experimentation is usually determined by the scientific problem to be investigated. For extrapolation to long-term effects at the population level more complicated functions will be needed. In particular, time-dependent effects will also have to be considered. There is a broad literature on such models that incorporates theoretical knowledge of the kinetics of substances, of growth, energy consumption and so on. However, our data do not always justify the fit of complex models to simple data sets. So, good advice is to use models which are parsimonious: as simple as possible and as complicated as necessary.

4.3.3.3   Analysis of quantitative response data sets

We do not go into the technical detail of model fitting here. The fitting criterion we use is the familiar least squares. The examples serve to illustrate the use of regression models for the investigation of ecotoxicity with measurement data. The interested reader is referred to any introductory text on regression analysis, such as the book of Draper and Smith (1981).

For the computation there is a wide range of statistical software available. Multiple linear regression methods are provided by almost all statistical packages. Special programs for non-linear regression are also available in the major statistical software (SAS, S-Plus, BMDP, GENSTAT, etc.).

4.3.3.3.1   An application of the parabola model

The data in Table 4.6 come from a 21-day reproduction test with *Daphnia magna*. The cumulative offspring during a three-week period is recorded for each of 10 daphnids per group exposed to different concentrations of 3,4-di-chloro-aniline (DCA). For these data, there is no special shape of the concentration–response curve. It can easily be fitted using a simple function such as the parabola. It is important to keep in mind, that such a merely

Table 4.6 Testing toxic effects on *Daphnia* reproduction. DCA concentration and number of offspring, $n = 10$ per group planned. Dots indicate missing values. (Data from Dr. Steinhäuser, Umweltbundesamt Berlin.)

Concentration	Number of offspring									
0.04	●	2	7	16	4	3	5	10	3	4
0.02	69	●	74	83	50	81	77	94	66	77
0.01	105	96	99	94	105	87	69	64	101	91
0.005	97	105	89	80	61	75	93	89	96	81
0.0025	113	113	93	87	95	91	110	87	102	120
0	87	92	97	99	104	103	111	82	86	89

descriptive model must not be used for an extrapolation beyond the experimental range. It can be fully sufficient for an interpolation within the concentration range of the experimental data. We get the fit using linear least squares as performed by most of the statistical software packages. Figure 4.5 shows a concentration-dependent decrease in reproductivity. We see that this simple model gives a reasonable fit to the data. Using the estimated model

$$\mu(c) = 95.4 + (25.6 \times c) - (56344 \times c^2) \qquad (4.19)$$

we calculate EC_{10} with

$$\mu(EC_{10}) = (1 - 0.1) \times \mu(0) \qquad (4.20)$$

and get $EC_{10} = 0.013$. It is even possible to estimate a lower confidence limit for this value, the so-called benchmark concentration. With $\alpha = 0.05$ we get the benchmark concentration $BC_{10} = 0.009$.

If we compare the different concentration groups with the control, we find that NOEC = 0.01.

From our model we conclude that near $c = 0.01$ we expect approximately 10% reduction in reproduction. Labelling this concentration a NOEC might suggest that no effect at all on reproduction is expected. This is obviously not correct. Using the benchmark concentration BC_{10} we would be able to state in a statistically justified way, that below $BC_{10} = 0.009$ an effect of not more than 10% reduction in reproduction is expected, adding that this statement has an error risk of 5%.

4.3.3.3.2 A non-standard model

We take the data from Table 4.3 on the relative change in volume of algae when treated with a herbicide. With these data, having only two replications per concentration, the NOEC approach would not make sense.

A regression-based analysis appears more appropriate. From the plot of the data (see Figure 4.2) we conclude that neither a linear nor a polynomial function would be suitable to fit the sharp decrease in volume. There is also no

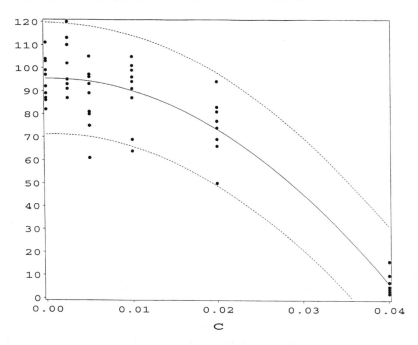

Figure 4.5 Concentration–response parabola fitted to the number of offspring until day 21 of *Daphnia magna* exposed to 3,4 di-chloro-aniline. 95% confidence interval for individual values included.

sigmoidal shape of the concentration-response curve using the original concentration values. Note that on a log-scale the sigmoid curve will appear, so an analysis using a logistic model would be appropriate.

In this example we consider another type, well known from biochemistry, the Michaelis–Menten model. Further, we transform our original values using the cube-root function, such that change in diameter instead of volume is modelled. This transformation gives more homogeneous residuals, such that the standard assumption of constant residual variance is better met.

The Michaelis–Menten equation can be written as

$$\mu(c) = R_{max} \times \{ 1 - c/(K + c)\} \tag{4.21}$$

where R_{max} denotes the maximum value at $c = 0$ and K corresponds to the EC_{50} value. A non-linear regression fit to the data gave $R_{max} = 4.68(0.1)$ and $K = 28.4(2.6)$ as parameter estimates. The iterative procedure was initialized with starting values for R_{max} and K. These can be found from a plot of the data as the maximum and a rough estimate of the EC_{50} value (for example R_{max}).

From the graph (Figure 4.6) we see that for a concentration $c = 28.4$, the model function is half the maximum value. The estimated value for R_{max} is

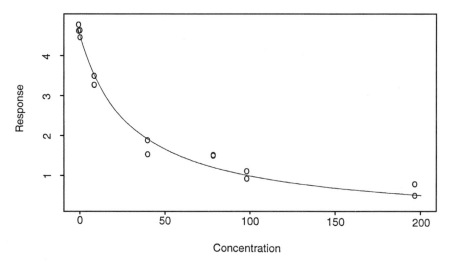

Figure 4.6 A Michaelis–Menten model fitted to the cube-root transform of relative change in volume of algae after treatment with a herbicide.

reasonable as $100^{(1/3)} = 4.64$ (Table 4.3). For the Michaelis–Menten function it is easy to calculate effective concentrations other than EC_{50}. After some algebra we obtain

$$EC_{100p} = K \times p/(1 - p) \tag{4.22}$$

so we calculate for our data an estimated 10% effective concentration as

$$EC_{10} = 28.4 \times 0.1/0.9 = 3.16 \tag{4.23}$$

We have to be careful in interpreting our results and have to keep in mind, that the model was built for the cube-root transform of volume. A word of caution – a 50% decrease in diameter gives a 87.5% reduction in volume.

4.3.4 NOEC OR BENCHMARKS – WHAT TO CHOOSE?

After having seen some useful approaches to describe the toxicity of a substance to our test organisms we are now going to discuss the merits and pitfalls of these approaches.

Progress in ecotoxicological research has been driven by legal requirements. Governments had to make decisions whether to allow the use and production of substances. So environmental concentrations had to be related to concentrations tested in experiments. If an environmental concentration had no influence on any organism it could be regarded as safe. This may be a reason for the predominant role of NOEC values in ecotoxicology. NOECs are easily calculated and communicated. But, the scientific community felt more and more uncomfortable with the NOEC approach. Some of the key arguments against the use of NOEC values for risk assessment are:

- NOECs are often misunderstood and misinterpreted: a non-significant result of a statistical test does not mean that there is no effect
- NOECs of good experiments tend to be smaller than those of inferior experiments
- NOECs depend heavily on the sample size and on the concentration pattern used.

From a statistical point of view, a replacement of the NOEC with a risk quantification that is unbiased, has a precision regularly depending on the sample size, and is not restricted to the experimentally tested concentrations is desirable.

The alternative is available, EC_{100p} values along with confidence bounds have all the desired properties. The benchmarks, lower confidence bounds for EC_{100p} values as proposed by Crump (1984) can be derived for quantitative as well as for quantal response data. A good choice for p depends on the importance of an effect and on the coefficient of variation. More work is needed to give an unambiguous choice.

4.4 DISCUSSION AND FURTHER READING

Investigating the toxic potential of compounds in the environment using the results of laboratory experiments is a challenge. The fit of models to quantitative and to quantal response plays a major role in this exercise. Modern statistics offers many techniques often already canned into statistical software packages that enable even non-specialists to perform the computational work necessary. For quantal-response data the book of Morgan (1992) is a rich source for modelling, and extends the standard models for the treatment of non-standard data. It also contains a comprehensive guide to the literature. The basics of regression analysis can be found in many textbooks. Draper and Smith (1981) can be considered as a standard reference. Techniques for non-linear regression can be found in Bates and Watts (1988) or Seeber and Wild (1989).

Special textbooks covering methods and theory for statistical models in ecotoxicology are Kooijman (1993) and Newman (1995).

Of course, statistical modelling can only be regarded as one step in the process of ecotoxicological risk assessment. But data analysis deserves effort and attention to avoid expensive experimental information being wasted.

REFERENCES

Bates DM and Watts DG (1988) *Nonlinear Regression Analysis and its Applications.* Wiley, New York.

Bliss CI (1935) The calculation of the dosage-mortality curve. *Annals of Applied Biology*, **22**, 134–167.

Crump KS (1984) A new method for determining allowable daily intakes. *Fundamental Applications of Toxicology*, **4**, 854–871.

Draper NR and Smith H (1981) *Applied Regression Analysis*, 2nd edn, Wiley, New York.

Kooijman SALM (1993) *Dynamical Energy Budgets in Biological Systems: Theory and Application in Ecotoxicology*, Cambridge University Press, Cambridge, UK.

Morgan BJT (1992) *Analysis of Quantal Response Data*, Chapman and Hall, London.

Newman MC (1995) *Quantitative Methods in Aquatic Ecotoxicology*, Lewis Publishers, Boca Raton, FL.

OECD (1993) *OECD Guidelines for the Testing of Chemicals*, OECD, Paris, France.

Rand GM and Petrocelli SR (1985) *Fundamentals of Aquatic Toxicology*, Hemisphere Publishing Corporation, New York.

Seeber, GAF and Wild, CJ (1989) *Nonlinear Regression*, Wiley, New York.

Analysis of Field Studies: Regression Analysis

ERIC P. SMITH AND CHRISTINE ANDERSON-COOK
Department of Statistics, Virginia Tech, Blacksburg, VA 24061-0439, USA

5.1 INTRODUCTION

Many of the data analysis problems encountered in ecotoxicology search for relationships between ecologically meaningful parameters and environmental measurements. Regression is a common statistical tool used for investigating, modelling and describing these relationships. In the broadest view, regression analysis includes one- and two-sample testing, analysis of variance, many multivariate methods as well as the analysis of association between sets of variables. In this chapter, the focus is on the common view of regression analysis, which includes simple linear regression and multiple regression. Additionally, the analysis of binary and other types of non-normal data will be discussed. Output and graphical results from each of the worked examples are given from the statistical computer package MINITAB (1996). Most statistical packages available on the market today have the ability to perform simple and multiple regression, although output format will vary slightly. For some of the more specialized operations such as variable selection and generalized linear models, other packages may be required.

Regression analysis is the analysis of association and potential causation. The method is a useful technique for:

1. Describing and modelling the relationship between a variable of interest (the dependent or response variable) and one or more variables (the explanatory or independent variables).
2. Predicting a value of the response for some particular values of other variables.
3. Testing hypotheses about associations between variables.
4. Comparing alternative models of association.

Examples of the use of regression analysis are abundant in the ecotoxicological literature. A perusal of one issue of *Environmental Toxicology and Chemistry*

Statistics in Ecotoxicology. Edited by T. Sparks. © 2000 John Wiley & Sons Ltd

indicated at least three articles using regression analysis. Elliot *et al.* (1996) used regression analysis to study relationships between concentrations of chlorinated polycarbons and effects in bald eagles. As part of a study the effect of nutrition on the survival and reproduction of Ceriodaphnia, Ferrari and Ferard (1996) used regression to relate the number of young produced per adult female to the coefficient of variation in reproduction (measured between beakers receiving the same nutritional supplement), showing that as the number of young produced increased, the coefficient of variation decreased. Hence, higher reproduction improved interbeaker reproducibility. Finally, Rattner *et al.* (1996) used regression analysis to study the relationship between hepatic microsomal activity in pipping herons and total PCB burdens.

5.1.1   THE DATA ANALYSIS APPROACH

The analysis of data using regression methods follows the standard approach of data analysis:

1. Preliminary evaluation of the data and relationship. The first step typically is graphical, since plots of data are useful for looking for pattern and anomalies. In regression analysis, the patterns of interest are related to the model considered. Typically, linear relationships and deviations from linearity are key features of the data we seek to identify.
2. Modelling the data. Models for data may be simple empirical models or more complex biologically motivated models. The choice of the model depends on the purpose of the analysis, the characteristics of the data, and understanding of the mechanism connecting variables.
3. Checking the model fit and assumptions.
4. Analysis of the results of the fitted model. This step involves testing of the model, its fit, as well as interpretation of values obtained by the model.

   The remainder of the chapter presents examples with discussion for simple linear regression, transformation of variables, multiple linear regression with model selection strategies and logistic regression for binary data. This represents only a subsection of the possible model types that might be encountered with regression applications from ecotoxicological studies, but should provide an overview of the general types of considerations that will be common to many of these problems. General references such as Myers (1990), Collett (1991) and Kleinbaum (1994) should be consulted for additional details.

5.2   SIMPLE LINEAR REGRESSION EXAMPLE – SULPHATE LEVELS AND TREE AGE

5.2.1   INITIAL EVALUATION OF DATA

The simplest case to consider involves a single explanatory variable and an associated response. For example, a study of 25 Welsh catchments considers

Figure 5.1 Scatterplot of sulphate versus tree age for 25 Welsh catchments.

the relationships between the ageing of conifer plantations and various chemical levels in soilwater (Appendix 4). The explanatory variable is tree age (measured in years), and it is thought to influence the level of sulphate (in mg l^{-1}) in the soilwater. Figure 5.1 provides a graph of sulphate versus tree age and is called a scatterplot. Scatterplots are useful for observing pattern, linear and non-linear trends, unusual observations and gaps in the data. Examining Figure 5.1, there does appear to be an association between the two measures. Note that several observations are extreme in that they fall either somewhat above or below the general trend, which may indicate that these observations do not follow the pattern of the remainder of the data as closely. Another special feature of the data is that there are several observations where tree age is equal to zero, which represents unplanted forest.

A preliminary measure of the level of association between the response and the explanatory variable is called the correlation. It is calculated as the standardized sum of terms for each observation, where each term represents the product of how much the one variable, x, is above or below its mean multiplied by how much the other variable, y, is above or below its mean. The sum is standardized by dividing by the standard deviation of both x and y (s.d.(x) and s.d.(y)), to scale the correlation so that it lies in the interval $[-1, 1]$.

$$r = \frac{\frac{1}{n-1} \sum_{i=1}^{n} (x_i - \bar{x})(y_i - \bar{y})}{s.d.(x) \, s.d.(y)} \tag{5.1}$$

The magnitude of the correlation gives an indication of the strength of the relationship between components as 0 implies no relationship and exactly 1 or -1 is an exact linear relationship between components. A negative correlation implies that as x increases, y decreases and a positive value implies that x and y increase together. For this data, r is calculated as 0.84 which is a large positive correlation, suggesting a strong relationship between the two variables.

5.2.2 MODELLING THE DATA

The simplest and perhaps most common model fitted to data is the simple linear regression model. There are two ways to describe the linear relationship. These are the standard form: $ay + bx = c$ and the slope intercept form: $y = mx + b$. In regression analysis, the slope intercept form is preferred as it focuses on y as being related to or caused by x. In this equation, m, the slope is viewed as the effect on y of changing x by one unit. The intercept, given by b, represents the value of y when x has a value of zero. To fit the model, the parameters associated with the straight line must be estimated. The eye fit tries to make the line go through the cloud of points, being as close as possible to all the points. Figure 5.2 graphs one such possible line through the data. Choosing a line specifies certain parameter estimates.

Fitting by eye is unsatisfactory as different eyes produce different fits and a common method for selecting parameter estimates is the least squares method. Least squares is a mathematical approach which mimics the process of fitting by eye without the subjectivity. The approach is based on trying to make the uncertainty about the line as small as possible. In statistics, uncertainty is measured using variation, typically about the mean value. In regression analysis the mean of the response is not fixed, but changes with the explanatory (x) variable (in a linear fashion). Thus, the difference between the response value and its fitted value (the estimated mean) measures uncertainty. Figure 5.2

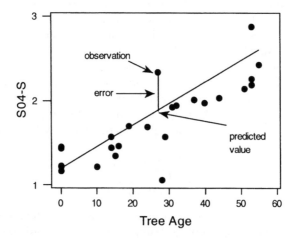

Figure 5.2   Scatterplot with line fitted by eye and deviation from the fit indicated by the error.

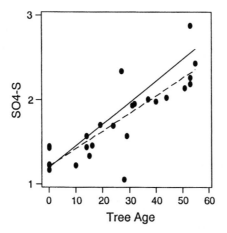

Figure 5.3   Scatterplot with line fitted by eye (solid) and regression line (dashed).

illustrates this uncertainty as error for one observed value, with its matching predicted value from the model.

When all the error values are squared then averaged together the result is proportional to variance. If the squared values are simply summed, the term is called a sum of squares error. The least squares approach is to pick the coefficients for the regression by making the variance or uncertainty about the regression line small. Figure 5.3 displays two lines used to describe a dataset. The solid line is fitted by eye. Contrast this with the line found with least squares to minimize variance, which is given by the dashed line. The sum of squared errors will be smaller for the line fitted by least squares than for the line fitted by eye.

The least-squares method is a preferred method for fitting data. First, the least squares method is closely connected to variance and making the uncertainty about the regression fit small. Second, the least squares method is easily extended to more complex situations, namely the case where we want to fit more than one variable where visualizing the data are more difficult. Finally, when the distribution associated with the data is normal, statistical inference (tests and confidence intervals) is based on variance and sum of squares. The least squares method provides the appropriate measures for making these inferences.

Table 5.1 displays computer output for the regression analysis using SO_4S as the response variable and tree age as the explanatory variable. Standard to most output is information about the parameter estimates, the significance and importance of the model and the significance of the parameters in the model. The regression model is estimated with slope equal to 0.0208 and intercept equal to 1.21. Thus a change in age of one year is associated with a change in SO_4S of 0.0208. The intercept describes the SO_4S content in unplanted forests.

Table 5.1 Regression

```
The regression equation is
y = 1.21 + 0.0208 Tree Age
Predictor      Coef          StDev          T          P
Constant       1.21073       0.08795        13.77      0.000
Tree Age       0.020809      0.002770       7.51       0.000

S = 0.2564     R–Sq = 71.0%    R–Sq(adj) = 69.8%


Analysis of Variance

Source         DF    SS        MS        F         P
Regression     1     3.7091    3.7091    56.44     0.000
Error          23    1.5115    0.0657
Total          24    5.2207


Unusual Observations

Obs    Tree Age    SO 4–S    Fit      StDev Fit    Residual    St Resid
13     27.0        2.3400    1.7726   0.0514       0.5674      2.26R
14     28.0        1.0600    1.7934   0.0516       −0.7334     −2.92R
24     53.0        2.8800    2.3136   0.0911       0.5664      2.36R

R denotes an observation with a large standardized residual
```

The analysis in Table 5.1 also provides information about the importance of the entire model and specific parameters within it. Evaluation of the model can be approached in two ways: as tests on individual coefficients and as an overall test on the model. The tests on the coefficients are carried out using t-tests. These tests evaluate the hypothesis that the parameter is equal to zero. For example, the simple linear regression model has two parameters, the intercept and the slope. In the table, the intercept is referred to as the 'Constant' and slope as 'Tree Age', which refers to the explanatory variable in the model. The test of the hypothesis that the intercept is equal to zero is based on

$$t = \frac{\text{estimate} - \text{hypothesized value}}{\text{estimated standard error}} \tag{5.2}$$

where the hypothesized value is set to zero. In this case, the estimate is 1.21073 and the estimated standard error is 0.08795 so the resulting t-test is 13.77. The table also provides a p-value for the test which is 0.000. It should be noted that the p-value is not exactly equal to zero, as indicated by the printout, but rather that it is rounded to zero when considering the result to three decimal places. The p-value provided is a two-sided p-value and the alternate hypothesis that is tested is that the coefficient in not equal to zero. Based on the size of the test statistic (13.77) which is very large relative to plausible values from a t-distribution or the size of the p-value, which is very small, we would reject the null hypothesis that the intercept is equal to zero. This means that the data observed are not consistent with the hypothesis that sulphate is equal to zero for catchments not planted with trees. If a different hypothesized value were of interest, the same test statistic equation listed above would be used, with the new value substituted into the equation.

The test on the slope is done in a similar fashion. The hypothesis evaluated in the table is that the slope is equal to zero versus the alternate that the slope is not equal to zero. In this case, the t-statistic is 7.51 which is large and the associated p-value is 0.000 which is very small. We would then reject the null hypothesis and conclude there is evidence of a relationship between sulphate and tree age. The slope parameter being equal to zero is an alternate way of saying that there is no linear relationship between variables, since if slope were zero, we would have the same predicted value of sulphate for all values of tree age. If the slope were zero, tree age would not be useful in predicting sulphate.

The significance of the slope may also be evaluated using an analysis of variance table. The interpretation of the table information is as follows. The total sum of squares measures how different the observations are from the overall sulphate mean. Some of these differences are due to the linear relationship between the variables. The regression sum of squares is the amount of the differences which may be attributed to the regression or which is 'explained' by the regression. The error sum of squares accounts for that which is 'unexplained' by the regression. These terms sum to the total. A commonly used summary statistic from regression is R^2 or the coefficient of determination. R^2 is defined as the ratio of the sum of squares for regression divided by the total sum of squares. It is thus the proportion of variation which may be attributed to the regression. R^2 is a number between 0 and 1. When there is no linear relationship, the value is near zero. If there is a strong linear relationship, the magnitude of the value is closer to 100%. For the catchment sulphate data, R^2 is 0.710 or 71.0% which suggests a strong linear relationship. Thus, the interpretation is that 71% of the variation may be explained through the linear relationship between sulphate and tree age.

The adjusted R^2 of 69.8% gives an estimate of what percentage of the total variation is explained by the model for other observations from the overall population of Welsh catchments. The adjusted R^2 is always less than or equal to the observed R^2, since it adjusts for the fact that the parameter estimates are the best for the observed sample but may not be ideal for future samples.

The other parts of the analysis of variance table include the degrees of freedom and mean squares. Mean squares are sum of squares divided by degrees of freedom. The degrees of freedom reflect the information required to estimate the sum of squares. The total degrees of freedom are $n - 1$ as there are n observations and the total (corrected) sum of squares requires estimation of the mean. The degrees of freedom for the regression sum of squares is 1 as it is based on a single value, the slope estimate. The degrees of freedom for error are $n - 2$, reflecting the number of observations adjusted for estimation of the slope and intercept. The mean squares are variance estimates. These estimates should be close in value when there is no linear relationship, but the mean square for regression should be larger than the mean square for error if there is a linear relationship. The ratio of these mean squares provides a test of the relationship and is an F-statistic. The significance of the relationship is evaluated using the p-value. For the example data, the F-statistic is large (56.44) relative to values commonly obtained from an F-distribution with 1 and 23 degrees of freedom and significant (the p-value is 0.000). Note that this is the same p-value for the test on the slope and in fact these two tests are equivalent and test the same hypothesis of no relationship between response and predictor variables. Also the square of the t-test on the slope is simply the F-statistic. The two approaches differ for multiple regression. In multiple regression, analysis of variance provides a useful way to compare models and evaluate components of a model while the t-tests provide a means of evaluating individual parameters.

5.2.3   CHECKING MODEL ASSUMPTIONS

Once a final model has been obtained to describe the relationship between the explanatory variable and response, it is important to check how adequately the model fits the observed data. Specifically, a number of assumptions were made in fitting the straight line to the data using least squares. The equation of the straight line relating x and y is modified to include a term which represents the inherent uncertainty of the data. Namely, $y_i = mx + b + e_i$, where y_i represents observation 'i' and e_i gives the amount that the observed value deviates from the predicted straight line. Figure 5.2 gave an example of an error term for a specific observation, if we were to assume that the true but unknown relationship is given by the straight line shown. The first assumption involving e_i, the error term, is that on average these values will not be consistently positive or consistently negative. This means that we expect the best straight line for the data will go through the middle of the observed data. If the error terms were consistently positive for a portion of the range of explanatory variable, for example, then this would imply that the best line would go below the observed data – not generally an intuitive model. The second assumption is that the variance of the error terms is the same across all observations. This implies that we expect the observations to have similar spread around the straight line.

Another assumption is that the data are independent of each other. This means that information about the value of one observation is not directly helpful in predicting another observation. When observations are not independent, variance is often smaller than if independent observations are used. This may be quite important as incorrect inferences may be made.

Finally, an additional assumption is sometimes made, namely that the error terms are normally distributed. This assumption is not nearly as critical as the first two for model fitting, but can be checked as well. While we are not able to

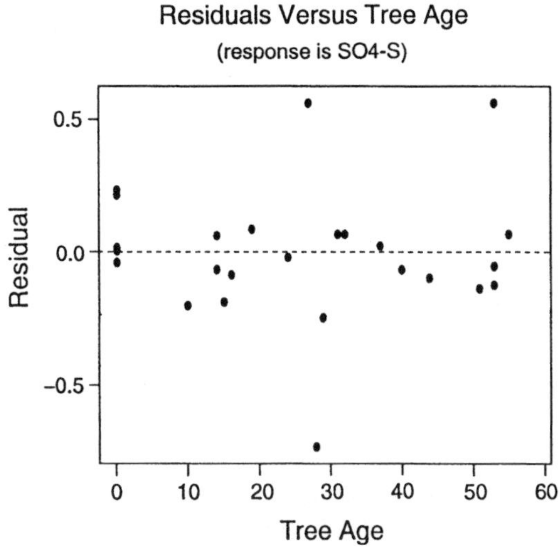

Residuals Versus Tree Age
(response is SO4-S)

Figure 5.4   Residuals versus tree age for Welsh catchment sulphate data.

calculate the actual error terms, we are able to estimate them by examining the difference between the observed values and the value suggested by our model. These values are called residuals, since they represent the part of the data that is left over after fitting the model.

Figure 5.4 shows a residual plot of the Welsh catchment sulphate data versus the explanatory variable, tree age. To evaluate if the first assumption is reasonable, we look for any strong patterns in the residuals that indicate some portion of the range might have a non-zero average value. Here the observations seem to be evenly scattered around the zero value for all values of tree age, which is consistent with the first assumption being true. We can also use this plot to consider the second assumption of uniform variance. In this case we are looking for the spread of the observations around zero being similar across the range of tree ages. It should be noted that only strong departures from the assumptions should be cause for concern, as all datasets will have some small patterns in the data. Therefore, for this data set the first two assumptions seem adequately satisfied. The assumption of independence cannot be checked directly with a plot, but must be assessed based on how the data were collected. The third assumption of normality can be evaluated by looking at a plot of the ordered residuals compared with the quantiles of a normal distribution. If this plot looks close to a straight line, then the residuals are reasonably assumed to be normal. Figure 5.5 shows the normal probability plot of the residuals. In this case, since the line is quite curved at the ends with three residuals (13, 14 and 24) well away from the rest of the line, we would be better off to not assume normality.

An interesting note is that the regression software in Table 5.1 identified

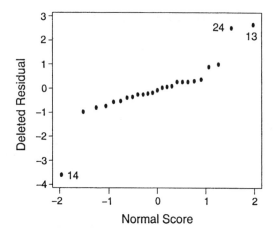

Figure 5.5 Normal probability plot of residuals for catchment data. Observation numbers were added to identify extreme residuals.

three observations (13, 14, 24) that have large residuals. These three are easily identified in all of the plots of the residuals as they are much further away from zero than the remaining data. It is recommended for these observations, that some time is spent trying to determine if there are some inherent difference between these and the remaining observations which might explain their lack of fit to the model.

5.2.4   INTERPRETING THE MODEL

Now that we have checked that the assumptions of the model are adequately satisfied, we are able to use the model for prediction. First, we may be interested in interpreting the coefficients of our model. For example, our model predicts that for each additional year that the trees age in a catchment, the level of sulphate will increase 0.0208 mg l^{-1} based on the estimate of the slope. If we wanted to attach a bound on the accuracy of the estimate, we could construct a 95% confidence interval for the slope, using the formula

slope estimate $\pm$ (tabulated t-value) $\times$ standard error of slope estimate   (5.3)

where the tabulated t-value is based on the t-distribution with the same number of degrees of freedom as the error sum of squares. Therefore for this example, a 95% confidence interval for the slope is

$$0.020809 \pm 2.069 (0.002770) = (0.0151, 0.0265) \qquad (5.4)$$

which means that the expected change from year to year of sulphate levels is between 0.015 and 0.027 mg l^{-1} for these types of catchments. It is important to

Table 5.2 Prediction for tree age = 20 years

Fit	StDev Fit	95.0% CI	95.0% PI
1.6269	0.0537	(1.5157, 1.7381)	(1.0849, 2.1689)

note that, at this stage, no assertion has been made about the age of the forest causing the sulphate to change, but rather that there is an association between the two measures. In order to establish causation between the explanatory variable, X, and the response, Y, a number of other possible explanations need to be eliminated: Y causes X, or Z, some other factor, causes both X and Y. It is frequently possible to justify that the first of these options is not possible, but to fully remove the possibility of another factor causing both X and Y usually requires that the process of data collection involves an experiment with random assignment of the levels of X to the experimental units. Frequently in eco-toxicological studies this is not possible or practical, and hence our conclusions are better summarized as an associative relationship.

Another common use of the model is to predict future observations for cases where we have values of the explanatory variable, but not the associated response. We can use the equation to give a single best predicted value for our new observation, and then obtain an estimate of the associated variation for that location as well. For example, if interest lies in a different catchment with trees of age 20 years, we predict that the level of sulphate would be 1.21073 + 0.020809 × 20 = 1.627 mg l^{-1}. We are also able to obtain a 95% prediction interval using MINITAB as shown in Table 5.2. Two intervals are provided: the first, a confidence interval, gives a range of possible values for the best fitting line. Its width reflects the uncertainty of our estimate of the actual line. The second is a prediction interval which has variation from the uncertainty of fitting the line as well as from our knowledge that future observations do not necessarily fall right on the line. Hence the prediction interval will always be wider than the confidence interval for the line.

5.3 TRANSFORMATONS – NO$_3$N AND TREE AGE

In some cases the initial plot of the data does not suggest a straight line relating the explanatory variables to the response. Several straightforward modifications to the procedure outlined above are possible to improve the fit of the model. Additional terms can be added to the model, such as quadratic or higher-order terms to help fit a curved relationship. A second possible option for improving the nature of the fit of the model to the observed data is to apply a transformation to either the response, or the explanatory variables, or both. Common choices are the natural logarithm, logarithm based 10 or the square root.

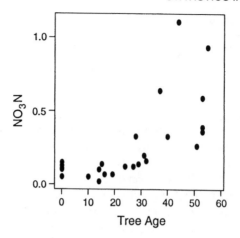

Figure 5.6 NO₃N versus tree age.

Consider an example from the catchment data where a new response of interest is considered. In this case, we are interested in modelling tree age to predict the level of the NO₃N (nitrate) in the soilwater. The response was transformed to give an improved fit for the model.

Notice how a straight line does not seem appropriate to summarize the relationship for the original data in Figure 5.6. In addition, we notice that the variability of the response becomes larger as tree age increases. In this case, we consider transforming the response using the natural logarithm. Figure 5.7 shows the new relationship.

With this transformation, the relationship now becomes more linear, and we have also reduced the change in the variance across the range of tree ages con-

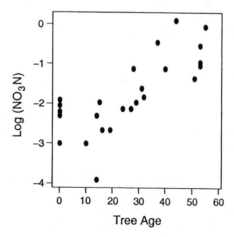

Figure 5.7 Log(NO₃N) versus tree age.

Table 5.3 Regression analysis output for model using tree age as the independent variable and log transformed NO_3N as the dependent variable

```
The regression equation is
Log (No3_N) = - 2.76  0.0387 Treeage

Predictor      Coef         StDev        T          P
Constant      -2.7610       0.2211      -12.49     0.000
Treeage        0.038727     0.006963      5.56     0.000

S = 0.6444    R–Sq = 57.4%    R–Sq(adj) = 55.5%


Analysis of Variance

Source        DF    SS       MS        F        P
Regression    1     12.847   12.847    30.93    0.000
Error         23     9.552    0.415
Total         24    22.399
```

sidered. In many situations where the variation in the response is proportional to the size of the response itself, the logtransformation has the effect of stabilizing the variance and making it more constant across the new values of the observations. Hence, the regression output is as given in Table 5.3.

To transform the data back to the original form for the response, the relationship becomes

$$\ln(y) = \beta_0 + \beta_1 X + \varepsilon$$
$$y = e^{\beta_0 + \beta_1 X + \varepsilon} \quad (5.5)$$
$$y = \beta_0^* \cdot \beta_1^{*X} \cdot \varepsilon^*$$

where we can redefine the parameters to be estimated. Hence this implies that the relationship between the response and the explanatory variable is not linear but rather of the form shown, and that the error structure for the model is multiplicative rather than the additive error studied in the previous example. To predict the level of NO_3N for 20-year-old trees, we would have to transform the prediction interval back to the original scale. We would calculate the value of $\log(NO_3N)$ as $-2.7610 + 0.03873(20) = -1.986$.

Fit	StDev Fit	95.0% CI	95.0% PI
−1.986	0.135	(−2.266, −1.707)	(−3.349, −0.624)

Therefore the predictions interval would be $(e^{-3.349}, e^{-0.624}) = (0.035, 0.536)$, with our best estimate of the value of NO_3N as $e^{-1.986} = 0.137$. The ability to transform either the response or explanatory variables provides much greater flexibility in model fitting and allows simple linear regression to handle a large and diverse group of applications.

Newman (1991, 1993) notes that the transformed estimate will be biased and suggests multiplying the estimated value by $e^{MSE/2}$ where MSE is the mean squared error from the regression model. In the above example, the correction factor is 1.23 and the corrected interval would be (0.043, 0.659). The bias correction factor depends on the value of the mean square error and is more important when variance is great.

5.4 MULTIPLE REGRESSION AND VARIABLE SELECTION – MERCURY CONCENTRATION IN FISH

Often in field studies involving the study of associations, there are multiple independent variables. For example, mercury concentrations in fish found in Florida lakes are an environmental concern. Lange *et al.* (1993) investigated the relationship between mercury concentrations and a variety of variables. In this study, four lake variables were used: total phosphorus, pH, alkalinity and chlorophyll *a*.

Multiple regression analysis is inherently more complex than simple linear regression. Visual display of pattern is complicated by inability to display many variables simply. Interpretation is more difficult because explanatory variables may be related to each other (jointly increasing or decreasing) and hence the effect of one cannot be evaluated without recognising how the others change.

The multiple regression model relates a single dependent variable with two or more explanatory variables in a linear fashion. If there are p explanatory variables that are of interest, the general form of the multiple regression model is

$$y = \beta_0 + \beta_1 x_2 + \beta_2 x_2 + \ldots + \beta_p x_p \tag{5.6}$$

where y is the variable of interest (dependent or response variable), x_i is the ith explanatory variable, β_0 is the intercept and β_i is the slope associated with the ith explanatory variable.

The model is flexible as explanatory variables may or may not be transformed prior to analysis. Thus, for example, if the explanatory variables are pH, N (nitrogen) and TP (total phosphorus) then the following models are all examples of multiple regression models:

$$\gamma = \beta_0 + \beta_1 \cdot pH + \beta_2 \cdot N + \beta_3 \cdot TP \tag{5.7}$$

$$\gamma = \beta_0 + \beta_1 N + \beta_2 N^2 + \beta_3 N^3 \tag{5.8}$$

$$\gamma = \beta_0 + \beta_1 \log N + \beta_2 \log TP \tag{5.9}$$

where β_0, β_1, β_2 and β_3 are parameters in the model.

Much of the discussion on testing and summary statistics for simple linear regression applies to multiple regression. In particular, a good model will be one with significant parameter estimates, a high proportion of variance explained (high R^2), low variance (small MSE) and good predictive ability. What is different about multiple regression is that the explanatory variables may be related to each other and this affects the values of the coefficients and hence model interpretation and relevant statistics.

5.4.1 VARIABLE SELECTION

The multiple regression model gives great flexibility in modelling relationships between variables. The flexibility leads to difficulties as choices must be made about which independent variables, including transformed variables, to include and how to relate the explanatory variables to the response. An important part of multiple regression analysis is the selection of a model or group of candidate models. Often in ecotoxicological studies, the number of explanatory variables is large and an approach must be used to select an appropriate set of variables and hence a 'good' model. The overall goal of model selection is to balance adequacy of fit and simplicity. Typically adding more variables can improve the ability of the model to explain the changes in the response, but this comes at the cost of having to collect more data. Additionally, adding superfluous variables may give the appearance of improved fit, but these variables may just be describing idiosyncrasies of the data collected, not the true underlying relationship. Hence smaller models with only significant variables frequently give better prediction for new data.

Characteristics of a good model include:

1. High explanatory power relative to other models.
2. Important variables (significant coefficients).
3. Good predictive performance.

The three criteria leads to three different approaches:

1. Select a model that results in the best value of some measure of explanatory power. This approach involves choosing the criterion, calculating the criteria for all possible (or many) models and selecting the optimal one. Possible criteria include:
   - largest R^2 (the coefficient of determination)
   - largest R^2 adjusted for the number of variables which penalizes the larger models
   - smallest mean square error or MSE
   - the C_p statistic, a statistic which is based on variance estimates, the number of variables and number of observations.
   Computer packages will list models and one or more of the above indices for models with different combinations of variables.
2. Select models based on the significance of individual coefficients. This approach leads to the 'stepwise' method in which the goal is to select a model in which all variables are significant and the model adequately describes the changes in the response. The stepwise method starts by selecting a value for the critical α level or corresponding F-value (based on

the F-distribution) for deciding if a variable should or should not be included in the model. At each step it should be determined if any variables should be removed from the current model, or new variables added. First evaluate if a variable can be removed by calculating an F-statistic for each variable in the model. The variable with the smallest non-significant F is removed. Then consider adding a variable by computing the F-statistic associated with each variable not in the model (i.e. a variable is added to the model and its importance is calculated by its F-statistic, or equivalently, the square of its t-statistic). Based on the critical F consider the ones which are significant and select the most significant variable to add to the model. Continue with this process until no variables can be added or removed.

3. Select a model with high predictive power. For evaluating model perform-ance it is better to use a separate dataset that was not used to estimate the model parameter or use the PRESS statistic (PRESS refers to predictive sum of squares). This statistic is calculated by leaving one observation out of the regression fitting process, computing the model, then predicting the omit-ted observation. Each of the observations may be individually removed and predicted by the other observations. PRESS is then the sum of squared errors associated with these predicted observations. The PRESS criteria can then be compared for all possible regressions to find a best model with high predictive power.

Regardless of the method used, the variable selection process should be thought of as a method of screening variables to determine if they are helpful for predicting the response value and leading to a set of candidate models. Other criteria are important for narrowing the field to a final choice. These criteria include having valid model assumptions and meeting the objectives of the study.

5.4.2   EXAMPLE: MERCURY IN FISH AND LAKE CHARACTERISTICS

Lange *et al.* (1993) studied mercury concentrations in largemouth bass in 53 different Florida lakes and how concentrations varied with lake characteristics. Water samples were collected at two times and pH, chlorophyll *a*, alkalinity and calcium measured. The measurements were averaged to give a representative sample for each lake. Largemouth bass were sampled in each with abundances ranging from 4 to 44 individuals. Because of bioaccumulation of mercury, mercury levels were adjusted to that for an age three fish. In 10 of the 53 lakes, ages were not measured and the values are the average concentration. Interest is in finding a model which relates lake characteristics to the age standardized mercury concentration.

The output from the stepwise regression procedure for the mercury con-centration data is shown in Table 5.4. The first portion gives the results from forward selection and shows the progression of variables that were added to the

Table 5.4 Summary of stepwise regression analysis and best subsets regression

a. Stepwise regression. Columns provide summaries for each step

Step	1	2
Constant	0.7222	0.7476
Alkalinity	−0.00557	−0.00443
t-statistic	−5.76	−4.19
Chlorophyll a		−0.0029
t-statistic		−2.24
Standard deviation	0.266	0.256
R-square	39.43	44.96

b. Best subsets regression. X indicates the variable is included in the model

Vars	R−square	R−sq(adj)	C–p	Standard deviation	Chl a	Calcium	pH	Alkalinity
1	39.4	38.2	7.0	0.266				X
1	37.6	36.3	8.7	0.270			X	
2	45.0	42.8	3.9	0.256	X			X
2	44.8	42.6	4.0	0.256			X	X
3	47.0	43.8	3.9	0.254	X		X	X
3	46.2	42.9	4.7	0.256	X	X		X
4	48.1	43.7	5.0	0.254	X	X	X	X

null model (i.e. only the intercept term). For example, here the first term that was added was alkalinity, which resulted in the model

$$\text{mercury} = 0.722 - 0.00557 \,(\text{alkalinity}) + \varepsilon, \qquad (5.10)$$

which explains 39.43% of the variability of mercury concentration (R^2-value). In the second step, the variable which produces the largest improvement in the fit of the model is selected. In this case, chlorophyll a is added to give the final model

$$\text{mercury} = 0.7476 - 0.00443 \,(\text{alkalinity}) - 0.0029 (\text{chlorophyll } a) + \varepsilon, \qquad (5.11)$$

with an R^2-value of 44.96%. At each step, an evaluation is made about which variable would be the best to add, and whether its addition will significantly improve the model. In this case, adding either of the two remaining variables (pH or calcium) does not improve the model enough to justify their inclusion. Hence, forward selection suggests the final model with just alkalinity and chlorophyll a.

The second portion of the output gives a subset of all the possible models that might be considered as the best subsets of variables to summarize the data. In Table 5.4b, the best two models with a given number of explanatory variables are given, with summary information about their fit to the data. The R^2 and R^2-adjusted values show what proportion of the variability of the response has been explained by the model. The C–p-statistic is a penalized score that rewards both good fit and parsimonious models. It is desired to have $C–p$-values less than or close to the number of parameters (variables plus one) in the model. Both $C–p$ and the standard error values indicate good models with small values. Notice that pH is the second-best variable for explaining the mercury concentrations in the fish; however, it is not included in the best model with two variables. This shows one of the key features of multiple linear regression. Since the explanatory variables are themselves interdependent, it is not always easy to predict which combinations of variables will do the best job explaining the response based on their individual associations with the response. Here alkalinity and pH are correlated, and explain some of the same features of the response, and so including chlorophyll a as the second variable gives a greater improvement to the model.

The output in Table 5.5 shows the stepwise procedure for the response transformed by the natural logarithm. Note the substantial improvement in the fit of the data, as measured by the R^2-value, for the transformed data. In

Table 5.5 Summary statistics for stepwise and best subsets regression for log transformed mercury concentrations

(a) Stepwise regression

Step	1	2
Constant	−0.4670	−0.2207
Chl a	−0.0199	−0.0138
t-value	−8.13	−6.40
Alkalinity		−0.0103
t-value		−5.94
S	0.544	0.421
R–sq	56.45	74.48

(b) Best subsets regression

Vars	R-Square	R–Sq (adj)	C–p	S	c h l a	c a l c i u m	p h	a l k a l i n i t y
1	56.4	55.6	36.3	0.54425	X			
1	53.6	52.6	41.9	0.56203				X
2	**74.5**	**73.5**	**3.0**	**0.42079**	**X**			**X**
2	64.1	62.7	23.2	0.49894	X	X		
3	75.5	74.0	3.0	0.41670	X	X		X
3	74.5	72.9	5.0	0.42506	X		X	X
4	75.5	73.4	5.0	0.42095	X	X	X	X

addition, the single best variable for explaining mercury concentration is now chlorophyll *a*, not alkalinity. It is not uncommon that the transformation of variables will yield a substantial improvement in the fit of the model, and will highlight different relationships between the explanatory variables and the response. The two-variable model, with chlorophyll *a* and alkalinity, is considered in more detail in Table 5.6.

Note that the coefficients given in the regression analysis match those specified in the first portion of the stepwise output, as the same model is being considered. The regression analysis indicates two observations with large residuals, one (observation 40) is quite large. The residual plots (residuals versus fitted values and each of the explanatory variables) suggest no irregularities other than the odd observations. An additional analysis repeated the steps using the data set without observation 40. The model which resulted in the same two variables being selected had similar coefficients and residual plots showed no major irregularities.

Table 5.6 Regression analysis for log mercury analysis using chlorophyll *a* and alkalinity as predictors

```
The regression equation is

log merc = - 0.221 - 0.0138 chla - 0.0103 alkalin

Predictor      Coef        StDev        T         P
Constant       -0.22067    0.08352      -2.64     0.011
chla           -0.013797   0.002155     -6.40     0.000
alkalin        -0.010331   0.001738     -5.94     0.000

S = 0.4208     R-Sq = 74.5%    R-Sq(adj) = 73.5%

Analysis of Variance

Source        DF    SS        MS        F         P
Regression    2     25.834    12.917    72.95     0.000
Error         50    8.853     0.177
Total         52    34.688

Source        DF    Seq SS
chla          1     19.581
alkalin       1     6.253

Unusual Observations
Obs    chla   log merc   Fit       StDev Fit   Residual   St Resid
  3    128    -3.2189    -3.1892   0.2093      -0.0296    -0.08 X
 38    152    -3.2189    -2.8709   0.2730      -0.3480    -1.09 X
 40     20    -0.1165    -1.4030   0.1072       1.2865     3.16R
 52      3    -1.2730    -0.4353   0.0710      -0.8377    -2.02R

R denotes an observation with a large standardized residual
X denotes an observation whose X value gives it large influence.
```

This section has presented an overview of a large subject area. Regression texts, for example (Myers 1990) give greater details on the topic of model fitting, selection and diagnostics. The diagnostic plots discussed here are a small sample of what is available. Myers (1990) and other regression texts should be consulted for additional details.

5.5 LOGISTIC REGRESSION AND GENERALIZED LINEAR MODELS – PRESENCE OF FISH TAXA

The least squares multiple regression model is a useful method for analysing data which is continuous and whose statistical properties do not deviate much from the normal model. In cases where the data are continuous but not normal, transformations often provide an approach to allow the methods to be used successfully. When the data are not continuous, transformations may sometimes improve the fit. However, there are other methods available which are more effective. For data from several distributions an analysis based on 'generalized linear models' is preferred (Maul 1992; Kerri and Meadow 1996). Two situations, the logistic model and the Poisson model are commonly relevant to ecotoxicological work.

The logistic regression model is applicable in situations where the response is binary. For example, in the laboratory a toxicity test often will result in mortality. The response might be recorded as 1 if the animal is dead after a certain period and 0 if the animal survives. The interest is in the probability that the animal survives for a certain period and is dependent on the dose that is applied. Another example occurred with the National Acid Precipitation Assessment Programme (NAPAP) in the United States (Baker and Gallagher 1990). In this monitoring programme, over 1000 lakes in the United States were sampled and chemical measurements collected. In addition, the presence or absence of relevant fish taxa was evaluated through sampling with gill nets. The response variable is thus the presence or absence of a fish taxa which may be explained by the chemical measurements. Logistic regression may be used to model other data which involves occurrences of animals. For example, a common usage of logistic regression is in habitat suitability analysis in which a habitat may be evaluated in terms of its ability to support mammals or birds. The response variable is whether or not an animal lives in a particular habitat and this is modelled in terms of explanatory habitat variables.

5.5.1 MODEL AND ANALYSIS

Investigations into what controls the presence or absence of a species can often be described through a cumulative function. For example, in an acid rain study,

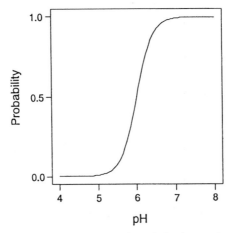

Figure 5.8   Cumulative distribution function used to mimic changes in survival probability as a function of pH.

it may be hypothesized that as pH increases, the chance that an individual is able to survive increases (at least over a reasonable range). Figure 5.8 illustrates this view. The statistical model that is used with this data has to satisfy two needs. First, it must be able to mimic the non-linear response to the chemical measurement, and second the model must account for the binary nature of the response (presence or absence). There are several models which may be used to mimic the S (sigmoidal) shape of the curve. The logistic model is a common choice, although the Probit model is also used in some applications. The logistic function is defined as

$$f(x) = 1/(1 + \exp(-x)) \tag{5.12}$$

For our presence–absence model, this may be rewritten as

$$\text{probability(present)} = \frac{1}{1 + e^{-x}} = \frac{e^x}{1 + e^x} \tag{5.13}$$

The plot of this function is given in Figure 5.9. The plots are identical except for scale. The model may be extended by replacing 'x' with a linear function. This leads to the linear logistic model

$$\text{probability(present)} = p = \frac{1}{1 + e^{-(\alpha + \beta x)}} \tag{5.14}$$

The model is non-linear but may be linearized using the logistic transformation:

$$\text{logit}(p) = \log(p/(1 - p)) = \alpha + \beta x \tag{5.15}$$

The second feature that must be accounted for is the probability model for

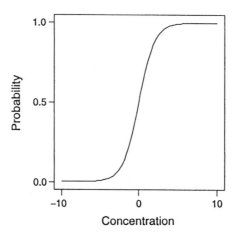

Figure 5.9   Plot of $1/(1 + e^{-x})$ versus x where x is concentration.

the data. The response variable can only have two values and is typically modelled with a simple model:

$$\text{probability (taxa present)} = p$$
$$\text{probability (taxa absent)} = 1 - p \tag{5.16}$$

where p is dependent on the x variable. The two pieces of the model may be combined to form the Bernoulli model for a variable Y which has values $y = 0$ or 1:

$$p(Y = y) = p^{y} (1 - p)^{1 - y} \tag{5.17}$$

Note that $y = 1$ results in the probability the taxa is present and $y = 0$ results in the probability the taxa is absent. The problem now is to estimate the parameters of the model. A complication is that p represents the probability which depends on the explanatory variables. The method used to estimate parameters in the logistic regression model is called iteratively reweighted least squares (IRLS) and is a computational method of obtaining ML estimates which works for numerous models. The method is based on the following idea: if the parameters are known, then $P(Y = y)$ can be computed. For a given set of observations $y_1, y_2, \ldots, y_n$, this can be computed for each observation. The information can then be combined to form the likelihood of the data. Given $P(Y = y_1), P(Y = y_2), \ldots, P(Y = y_n)$ this is computed as the product of the probabilities (and referred to using the letter L). With the ML method, the parameters are estimated to make the product of the probabilities as large as possible. It is common to compute the log of the likelihood ($\log L$) which is the sum of the log of the probabilities, as this results in a summary measure with good statistical properties.

To obtain the model estimates, the idea is to pick values for the parameters which maximize the likelihood and this is done using IRLS. This involves estimating the relationship, computing the estimate of p, computing the variance (weights) and re-estimating the relationship. Usually, the process converges and we obtain the estimates and summary statistics.

Part of the analysis involves computing the likelihood associated with the model. The likelihood is valuable for evaluating and comparing models. Given the data and parameters, the likelihood is computed by multiplying together the likelihoods associated with each observation. The result is a small number. Usually, this is not given on computer output, rather the log of the likelihood is computed. The log likelihood will be a negative number and small values are 'good'. With likelihoods, small is relative and the value of the likelihood is comparative.

As an example, consider the modelling of data on the presence and absence of brook trout in 413 lakes. The data come from a more extensive study (Baker and Gallagher 1990) in which roughly 50 variables were measured. The output in Table 5.7 overleaf considers the relationship between presence/absence and three variables. The log likelihood for this model is -230.021. This is used to compute the goodness of fit statistic by multiplying by -2 and is called the deviance goodness-of-fit test. This test is not used unless there is replication of the response at the different values of the explanatory variables, since otherwise the approximation is not very good.

5.5.2   LIKELIHOOD RATIO TESTS

Testing in generalized linear models is often based on the likelihood and related quantities. Since this measures how likely the data are given the parameters, different sets of parameters may be compared and evaluated based on how much change there is in the likelihood. For example, suppose there are two variables, say pH and acid neutralizing capacity (ANC). The model with both variables is fitted. To evaluate the importance of one variable, say pH, first compute the likelihood with both parameters. Then remove pH and recompute the likelihood. The ratio of the two likelihoods is called a likelihood ratio test. The test is actually computed by calculating -2 times the difference in the log likelihoods. This is used to compare two models, referred to as reduced and full. The distribution of this computed statistic is approximately chi-square with the number of degrees of freedom equal to the number of additional parameters in the full model.

In the acid rain example, the log likelihood for the model is -230.021 for the model containing the three variables. A test of interest is for the overall significance of the model. This test is evaluated by computing the log likelihood for the model without the variables (just containing a constant) and is -280.136. The negative of twice the difference between these quantities ($-2(-280.136 - (-230.021))$) is 100.230. This is referred to as a G-test in the output (Table 5.7)

Table 5.7 Logistic regression analysis of presence and absence of brook trout as a function of acid rain variables

Link function: logit

Response information

Variable	Value	Count
Brook trout	1	171
	0	242
	Total	413

413 cases were used
7 cases contained missing values

Logistic regression table

Predictor	Coefficient	Standard deviation	Z	P	Odds ratio	95% CI lower	95% CI upper
Constant	−0.5387	0.2243	−2.40	0.016			
Summer surface silica	0.2711	0.0487	5.57	0.000	1.31	1.19	1.44
Hydrogen-ion concentration	−0.0807	0.0188	−4.28	0.000	0.92	0.89	0.96
Acid neutralizing capacity	−0.00199	0.00063	−3.13	0.002	1.00	1.00	1.00

```
Log-Likelihood = -230.021
Test that all slopes are zero: G = 100.230, DF = 3, P-Value = 0.000

Goodness-of-Fit Tests

Method              Chi-Square   DF    P
Pearson              385.531    409   0.792
Deviance             460.043    409   0.041
Hosmer-Lemeshow        7.343      8   0.500

Table of Observed and Expected Frequencies:
(See Hosmer-Lemeshow test for the Pearson Chi-Square Statistic)
```

and is treated as a quantity from a chi-square distribution with three degrees of freedom.

A quantity related to the likelihood is called the deviance. The deviance of a model is defined as minus twice the difference of the log likelihoods of the model of interest and the saturated model. The saturated model is the model which has more parameters than are necessary and is used to represent the best attainable likelihood. As with likelihood, changes in the deviance as the model is altered are important for assessing the importance of the variables in the model.

Table 5.7 *Continued*

Value	1	2	3	4	5	Group 6	7	8	9	10	Total
1											
Obs	0	7	9	19	14	15	22	27	25	33	171
Exp	1.1	5.4	10.5	14.0	15.8	18.4	21.7	24.3	27.1	32.7	
0											
Obs	41	34	32	23	27	26	20	14	16	9	242
Exp	39.9	35.6	30.5	28.0	25.2	22.6	20.3	16.7	13.9	9.3	
Total	41	41	41	42	41	41	42	41	41	42	413

```
Measures of Association:

(Between the Response Variable and Predicted Probabilities)

Pairs          Number   Percent   Summary Measures
Concordant     31518     76.2%    Somers D                 0.53
Discordant      9733     23.5%    Goodman—Kurskal Gamma    0.53
Ties             131      0.3%    Kendalls Tau-a           0.26
Total          41382    100.0%
```

5.5.3 WALD AND SCORE TESTS

In multiple regression applications, it is common to test the significance of a parameter using a *t*-test. These tests are formed by dividing an estimator by its estimated standard error. With generalized linear models it is more common to use a *z*-test rather than a *t*-test and is referred to as a Wald test (see the first part of Table 5.7). The test is formed in the same manner, but is an approximate test whose validity depends on a reasonably large sample. In the example on acid rain, *z*-tests are given for the four parameters in the model. The *p*-values indicate that all are significant. The sample size is quite large hence the normal approximation should be reasonable.

Other packages may report a chi-square test or a score test. When the *z*-statistic is squared one obtains a chi-square statistic.

$$z^2 = \text{estimate }^2/\text{estimated variance} = \text{chi-square} \qquad (5.18)$$

The score test is an extension of this idea to the likelihood. The score is associated with the first derivative of the likelihood and the variance (called the information matrix) with the second derivative. Score tests differ slightly from Wald tests but generally are close numerically. With large samples there are often negligible differences. Score tests tend to be slightly preferred with smaller sample sizes.

Besides the test statistics, the output also includes odds ratios for the three variables in the model. The odds ratios are computed as the exponential of the coefficient. Thus, $\exp(-0.0807) = 0.92$. The confidence interval is just the exponential of the endpoints of the confidence interval for the coefficient (using a z interval). Odds ratios are viewed relative to 1.0. A value of 1.0 indicates that the variable has no effect on the probability of presence or absence. The odds are defined as the probability that $Y = 1$ relative to the probability that $Y = 0$ and the odds for a predictor indicate how a one unit change in the predictor affects the odds. For the example, this refers to the probability that a taxa is present relative to the probability that it is absent. The odds for the hydrogen-ion variable is 0.92 which indicates that a one unit increase in hydrogen-ion concentration decreases the probability that the taxa is present. The odds for silica are >1, indicating an increase in silica increases the probability of observing the taxa. Finally, the odds for ANC are essentially 1, indicating that a unit increase of this variable has little effect on the observation of the taxa, given the other variables in the model.

Output from most logistic regression packages also includes information about goodness-of-fit and summary information on measures of association. The goodness-of-fit statistics are intended to indicate if the model 'fits' the data adequately. The null hypothesis is that the model fits versus the alternative that the model is not adequate. A non-significant test is often used in model fitting to stop the modelling process. Some caution needs to be observed in using the tests as the statistics are approximate. When the data are binary (e.g. presence absence data), the deviance test does not provide a good approximation and should not be used. The Pearson chi-square test is a goodness-of-fit test which compares observed and expected values. The Hosmer–Lemeshow test is similar, although it partitions the interval from 0 to 1 into sections, evaluates the expected number of estimated probabilities falling into the sections, then computes the test.

The final part of the output gives several measures of association. Three measures are presented in the MINITAB output and are based on agreement as a function of concordance. Concordance is agreement in probability. Suppose two observations are selected which have different values. If the estimated probabilities from the logistic analysis are in the same order (the lower value with the zero and the higher value with the 1), then the observations are concordant. For the acid rain data, there are 171 observations labelled as 1 and 242 observations labelled 0, resulting in 171*242 or 41382 pairs. Of these 31 518 or 76% are concordant. The three measures given in Table 5.7 are based on the number of concordant and discordant pairs. The Goodman–Kurskal gamma measure, for example, is given by

$$D = \frac{(\text{concordant} - \text{discordant})}{(\text{concordant} + \text{discordant})} \tag{5.19}$$

Other measures are defined in slightly different manner, for details, see Agresti (1984). The measures are interpreted similar to correlations. Values range from

−1 to 1, with the sign indicating a negative or positive relationship and the size indicates the strength of the relationship.

5.5.4 OTHER OUTPUT

One can output various diagnostic statistics which can be used to assess the fit of the model and dependence on individual observations. These include 'leave-one-out' statistics and residual statistics. These can be plotted to indicate observations which are not well predicted. Also, there are plotting techniques available to assess the influence of observations on the analysis (for details, see Collett (1991).

Some computer packages report statistics which are useful for comparing and selecting models. Two of these are the Akaike Information Criteria and Schwartz's Criterion which are given by

$$-2 \log \text{likelihood} + 2(k + s) \quad \text{and} \quad -2 \log L + (k + s) \log N \qquad (5.20)$$

respectively, where k is the number of ordered responses, s the number of explanatory variables and N the number of observations.

These are sometimes useful for comparing models adjusting for the number of variables and smaller values are desired.

5.5.5 NOTES

Different computer packages have different approaches to summarizing the output of the analysis. Collett (1991) gives an overview of the different packages. For example, SAS has a stepwise logistic program which uses the score test to add variables and the Wald test to remove variables. If the data are matched, a different approach is needed. The appropriate method is to form differences and then to analyse the differences (see Collett 1991 or Kleinbaum 1994) for details.

Plots of information are important for checking the assumption of the analysis and evaluating the fit of the model. There are many possible plots that may be used in the logistic regression context (more than in multiple regression) and these are described in Collett (1991). For example, there are several ways to define residuals for the logistic regression problem including Pearson residuals, likelihood residuals and deviance residuals. Pearson residuals are defined as

$$\frac{y_i - \hat{p}_i}{\hat{p}_i(1 - \hat{p}_i)}. \qquad (5.21)$$

Figure 5.10 displays the plot of standardized Pearson residuals (these are the Pearson residuals further adjusted so their variance is 1. Values exceeding 2 or

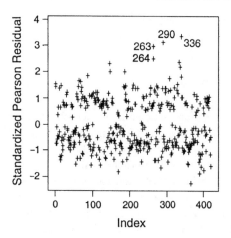

Figure 5.10   Plot of standardized Pearson residuals versus observation number.

-2 require some investigation as these residuals are viewed as large. The plot indicates several observations that are not well fitted. These correspond to lakes with fish present which according to the model should be absent. Removal of these observations did not strongly affect results.

5.5.6   OTHER MODELS

The non-linear approach used in logistic regression may be generalized to other distributions such as the Poisson distribution. The Poisson model may be applied in situations where counts are collected, for example, the abundance of a taxa at a site or reproductive data. For details see Dobson (1983).

5.6   CONCLUSIONS

Regression methods represent a powerful set of techniques for identifying key associations between responses and explanatory variables. Not only can they help provide understanding about the relationship between variables, but they can also be used to give empirical prediction models. Caution should be used in trying to extend a model from a description of association between components of the model to a causal relationship, since in general this can only be done through a designed experiment. Selecting the appropriate model and/or transformation of variables are key components of the model building process and should be done with care. Testing of model assumptions and adequacy of fit are also important parts of the modelling process. Once a good working model has been established, the researcher has a powerful tool for the description and understanding of an ecotoxicological process.

REFERENCES

Agresti A (1984) *Analysis of Ordinal Categorical Data*, Wiley, NewYork.

Baker J and Gallagher J (1990) Fish communities in Adirondack lakes. In *Adirondack Lakes Survey: An Interpretative Analysis of Fish Communities and Water Chemistry, 1984–87*, Adirondack Lakes Survey Corporation, Ray Brook, NY, pp. 3-1–3-11.

Collett D (1991) *Modelling Binary Data*, Chapman and Hall, London.

Dobson A (1983) *Introduction to Statistical Modelling*, Chapman & Hall, London.

Elliot, JE, Norstrom RJ, Lorenzen A, Hart LE, Philibert H, Kenedy SW, Stegeman JJ, Bellward G and Cheng KM (1996) Biological effects of polychlorinated dibenzo-*p*-dioxins, dibenzofurans and biphenyls in bald eagle (*Haliaeetus leucocephalus*) chicks. *Environmental Toxicology and Chemistry*, **15**, 782–793.

Ferrari B and Ferard J-F (1996) Effects of nutritional renewal frequency on survival and reproduction of *Ceriodaphnia dubia*. *Environmental Toxicology and Chemistry*, **15**, 765–770.

Kerri DR and Meadow JP (1996) Modelling dose response using generalised linear models. *Environmental Toxicology and Chemistry*, **15**, 395–401.

Kleinbaum, DG (1994) *Logistic Regression: A Self-Learning Text*, Springer-Verlag, New York.

Lange TR, Royals HE and Connor LL (1993) Influence of water chemistry on mercury concentration in largemouth bass from Florida lakes. *Transactions of the American Fisheries Society*, **122**, 74–84.

Maul A (1992) Application of generalised linear models to the analysis of toxicity data. *Environmental Monitoring and Assessment*, **23**, 153–163.

MINITAB (1996) MINITAB Release 11. MINITAB, Inc., State College, PA.

Myers RH (1990) *Classical and Modern Regression with Applications*, 2nd edn, Duxbury Press, Boston, MA.

Newman MC (1991) A statistical bias in the derivation of hardness-dependent metals criteria. *Environmental Toxicology and Chemistry*, **10**, 1295–1297.

Newman MC (1993) Regression analysis of log-transformed data: Statistical bias and its correction. *Environmental Toxicology and Chemistry*, **12**, 1129–1131.

Rattner, BA, Melancon MJ, Custer TW and Hothem RL (1996) Cytochrome P450 and contaminant concentrations in nestling black-crowned night-herons and their inter-relation with sibling embryos. *Environmental Toxicology and Chemistry*, **15**, 715–721.

6

Multivariate Techniques

ANDY SCOTT[1] AND RALPH CLARKE[2]

[1]NERC Institute of Terrestrial Ecology, Merlewood Research Centre, UK
[2]NERC Institute of Terrestrial Ecology, Furzebrook Research Centre, UK

6.1 INTRODUCTON

Most field surveys or observational studies record more than one, and often, many attributes about the objects of interest. For example, studies of the influence of potential pollutants on vegetation composition, which record the presence/absence or abundance of plant species within selected quadrats, often lead to a combined taxonomic list of over 100 species. Similarly, investigations of the toxic loadings present in wild animals usually record the concentrations of several chemical species in each sampled individual together with other characteristic features. The attributes recorded during such surveys can often be grouped into 'explanatory' variables, those potentially associated with the 'cause' of an environmental effect, and 'response' variables, that measure the 'effect' on the ecological community.

In statistical terms, the distinguishing feature of all such multi-attribute surveys is that they record more than one response variable. Each response variable could be analysed separately using the statistical procedures for univariate responses discussed in earlier chapters. The exploratory data analyses discussed in Chapter 2 can also help to build up an understanding of the variability and interrelationships within complex survey data. However, if there are a large number of attributes, with each analysed separately, the summary of results is often cumbersome, repetitive and difficult to synthesize. Moreover, important aspects of covariation and relationships between several response variables can be missed by analysing each variable in turn. Analysing all the responses together can often lead to more powerful statistical tests for the combined effect of pollutants and other environmental factors.

This chapter discusses the use of those statistical techniques, known collectively as multivariate analysis (MVA), which attempt to represent, synthesize or summarize the patterns of variation and interrelationships in the whole data set simultaneously. Multivariate techniques are distinguished from other

Statistics in Ecotoxicology. Edited by T. Sparks. © 2000 John Wiley & Sons Ltd

(univariate) statistical techniques by having as their main focus of interest a whole set of variables or measurements rather than a single variable. By this criterion, multiple regression (Chapter 5) is not regarded as a multivariate technique because it has only a single response variable, even though it involves more than one variable.

One initially surprising use of the multivariate techniques of this chapter is for the analysis of repeated measures data, whereby information on one variable is recorded at several time points for each of a sample of individuals. Because repeated measurements on the same individual are likely to be correlated, especially when taken at relatively short intervals, the observations are not independent. This correlation structure can be allowed for by treating the observations at the q points in time as q variables measured on the set of individuals and using the techniques of multivariate analysis. Other specialist forms of repeated measures analysis also exist; an example is profile analysis which assesses whether the changes between successive time points are the same for all groups of individuals (Seber 1984, section 3.6.4).

The development and presentation of multivariate analysis over the past 70 years has largely been in terms of matrix algebra, because the latter provides succinct, mathematically elegant expressions of relationships between any number of variables. However, many multivariate techniques can be explained, and are more easily understood, in graphical or geometrical terms. Since matrix algebra is a 'foreign' language to many potential users of MVA techniques, the mathematics of multivariate analysis are kept to a bare minimum in this chapter. Techniques are illustrated graphically and through the interpretation of worked examples of their application to ecotoxicology.

Adequate hardware and software are both needed for the use of a technique to develop. Until relatively recently the lack of computing power to manipulate large data matrices limited the spread of multivariate techniques in applied research. Moreover, the output from the software that existed was often expressed in mathematical terms and hence was difficult for applied users to interpret. In recent years hardware (in the form of the PC) has improved dramatically and software for most multivariate analyses has become readily available. Explanation of output has improved, especially through the use of graphical presentations of results. The result is that multivariate techniques are now well established in ecology and the aim here is to aid their expanding use in ecotoxicology.

It is not possible in a single chapter to provide anything like a comprehensive coverage of multivariate techniques or an adequate explanation of the details of each technique. Large numbers of statistical texts deal only with multivariate methods (e.g. Manly 1986) and many are devoted solely to individual techniques (e.g. Gordon 1981). The aim of this chapter is to give a brief, necessarily incomplete, overview of the subject while describing a few individual techniques in enough detail to provide an understanding of basic ideas and principles. We would urge the potential user of multivariate methods to refer to more specialist books before applying these methods. Current statistical software allows the easy application of many techniques which were previously difficult to use, but in doing so opens the way to many snares and pitfalls for the unwary user.

Many multivariate techniques have univariate analogues and these will be noted as each is discussed. MVA techniques can be classified in many ways. An important distinction is between those techniques which are primarily descriptive and those which are based on a distributional or parametric model for the data. Descriptive methods have in the past relied on subjective judgements for interpretation while parametric methods provide more formal tests of significance. However, increases in computing power are leading to a rapid growth in the use of randomization tests and resampling techniques (called 'bootstrapping') which can be used with non-parametric methods to formally assess potential effects and multivariate relationships (Manly 1997). In practice most parametric multivariate techniques make the assumption that the variables under consideration have a multivariate normal distribution and in using these methods the approximate validity of this assumption should always be examined.

There are many ways of classifying multivariate techniques. In this chapter we have chosen to use a division based on the objective of the techniques as follows:

1. Ordination or dimension reduction. Techniques which treat a data set as a whole and aim to reduce its complexity by expressing the information it contains in terms of a small number of new variables. Results are often expressed visually (section 6.3).
2. Methods for grouped data. Often data sets are obtained from a number of separate groups, for example different species, or are thought to contain such groups. Techniques for quantifying or assessing differences between predefined groups are discussed in section 6.4 and techniques aimed at uncovering groupings in a dataset (cluster analysis) are covered in section 6.5.
3. Relating two sets of variables. Often interest is focused on the relationships between two sets of variables; for example 'explanatory' and 'response' variables or environment and species data. Techniques for dealing with this situation are described in section 6.6.

6.2 PRELIMINARY DATA MANIPULATION: TRANSFORMATIONS AND SIMILARITY INDICES

6.2.1 STANDARDIZATION AND TRANSFORMATION OF VARIABLES

Because techniques for multivariate data analysis, by definition, involve several and often many variables, it is very easy to simply 'throw all the variables' into a computer package without giving careful thought to the varying scales of measurement or the statistical distributions of the individual variables. However, the results and interpretation from an analysis can depend critically on these factors.

As a simple example, if the chemical concentration of a particular solute is expressed in micrograms it will have 1000 times the standard deviation and 10^6 times the variance it would have if expressed in milligrams; this could give it an unwanted very high influence in some forms of multivariate analysis compared to other solutes measured in milligrams. If variables are naturally measured in completely different units (e.g. weight and age class) and it is desired to allow each variable, at least potentially, to contribute equally or be given the same importance in an analysis, then it is common to standardize each variable by subtracting its mean value from each observation and then dividing by its standard deviation. This gives each variable a mean of zero and a standard deviation (and hence variance) of one.

If the sample values for a variable have a skewed distribution with a few unusually large values, then these 'outliers' are likely to have a dominating influence on the total variation for that variable. Hence any analysis may tend to concentrate on 'explaining' these extreme observations. This problem is particularly important when using techniques which assume multivariate normality. In such cases it may be best to 'transform' the variable in some way before analysis and work with the transformed set of values. Measurements of chemical concentrations are particularly prone to this problem and taking the square root or logarithm of concentrations before analysis can greatly improve distributional properties. The use of pH rather than hydrogen-ion concentration is an example where such a transformation is already in common use. Such transformations can also be useful in analyses of grouped data when the within-group variability varies from group to group, since some multivariate techniques assume equal within-group variability (see section 6.4). It should always be remembered, however, that transforming variables can make the interpretation of analyses more difficult since results will be expressed in terms of the transformed variables and not the original measurements.

6.2.2 MEASURING SIMILARITY AND DISSIMILARITY BETWEEN SAMPLE UNITS

For some applications (such as cluster analysis, discussed in section 6.5) it is necessary or of interest to define how similar (or dissimilar) any two units in a data set are. Similarity can be measured in a wide variety of ways, but the choice of similarity measure depends on the types of variables involved (Box 6.1). Legendre and Legendre (1983) give an excellent account and comparison of many similarity measures. If a similarity index S can range from 0 to 1, then it can be converted to a dissimilarity measure by using $D = 1 - S$. Before calculating similarities it is often sensible to either transform the raw variables

Box 6.1 Commonly used measures of similarity (denoted by S) and dissimilarity (D) between a pair of units 1 and 2

(a) Binary variables (e.g. presence/absence of species)
Let a = number of species occurring in both units, b = number of species occurring in unit 1 but not in unit 2, c = number of species occurring in unit 2 but not in unit 1.

$S_1 = a/(a + b + c)$ Jacard
$S_2 = 2a/(2a + b + c)$ Czekanowski

(b) Quantitative variables (e.g. chemical concentrations or species abundance)
X_{kj} = value of variable j for unit k

$D_1 = \Sigma_j \mid X_{1j} - X_{2j} \mid$ City block metric
     = sum of differences in
     values regardless of sign
$D_2 = \{\Sigma_j \ (X_{1j} - X_{2j})^2\}^{1/2}$ Euclidean distance
$D_3 = (\Sigma_j \mid X_{1j} - X_{2j} \mid)/(\Sigma_j X_{1j} + \Sigma_j X_{2j})$ Bray-Curtis (1957)

(c) Any combination of types of variables

$S_3 = \Sigma_j C_{12j}/p$ Gower (1971)

where p is the number of variables involved, and where:
for a quantitative variable j, $C_{12j} = 1 - \mid X_{1j} - X_{2j} \mid/($range of variable $j)$;

for a qualitative variable j, $C_{12j} = 1$ if the units have the same state for the variable, and 0 otherwise; and for species presence/absence (ignoring jointly absent species) $C_{12j} = 1$ if both units 1 and 2 have species j, and 0 otherwise.

(e.g. use log or square root transform to downweight the influence of extreme values or occasional very high species abundances), or standardize to zero mean and unit variance (to eliminate dependence on the variable units of measurement and size of values of variables). With chemical concentrations it is perhaps best to measure dissimilarity by Euclidean distance (D_2 in Box 6.1) on the standardized variables. Clarke and Ainsworth (1993) recommend using the Bray–Curtis dissimilarity measure (D_3 in Box 6.1) for abundance or biomass data on marine taxa after using a double square root transformation to reduce the dominance of very abundant or large taxa respectively. This transformation has an advantage over the log transformation in that there is no need to add a constant to cope with zero values.

6.3 ORDINATION OR DIMENSION REDUCTION

As described in Chapter 2, an important first step in the analysis of any data set is the visual examination of the data. This provides an initial impression which

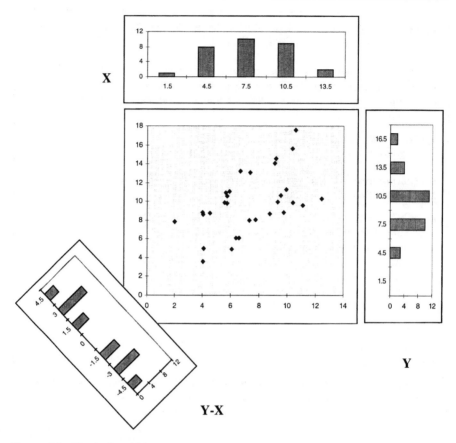

Figure 6.1 Marginal and bivariate distributions. Because of strong within-group correlation, groups in the data are not visible when variables *X* and *Y* are examined individually. The groups become obvious in the bivariate scattergram and from the derived variable *Y* − *X*.

can guide later, more formal analyses and may well reveal unsuspected structure in the data. With multivariate data each variable can be examined separately and scatterplots can be used to look at the relationship between pairs of variables. With modern computer software it is even possible to produce three-dimensional, rotatable plots to view three variables together. However, examining the interrelationships of more than three variables is not easy.

The solution to this problem is to realize that, geometrically at least, there is nothing special about our choice of variables. Visualize a three-dimensional scatterplot and imagine being able to rotate it in any direction. As the cloud of points rotates different images of the data come into view and some of these may represent the main features of the data better than others. Figure 6.1 gives a two-dimensional example of this, a data set where neither of the two original variables by themselves reveal that the points fall into two distinct groups,

---

Box 6.2 Calculating new variables

If there is a single set of q original variables called $X_1, X_2, \ldots, X_q$, then a derived variable V, calculated as

$$V = B_1X_1 + B_2X_2 + \ldots + B_qX_q,$$

is a linear combination of the original variables. The coefficients B_1 to B_q are numerical values chosen to give V some desired property, dependent upon the type of analysis.

If the original variables are transformed in deriving the new variable, e.g.

$$V = B_1\log(X_1) + B_2\log(X_2) + \ldots + B_q\log(X_q),$$

the result is a non-linear combination of the original variables.

---

groups which are obvious when looking at the data from a different direction. As well as rotating a cloud of data points we can imagine stretching or deforming it in order to emphasize the data structure.

Mathematically 'looking from a different direction' is equivalent to calculating a new variable which is a combination of the old variables, and transforming or scaling either the original or the derived variables is the mathematical equivalent of 'stretching'. Many multivariate techniques involve the derivation of such new variables, often referred to as axes (Box 6.2). The aim is to replace the original, usually large, set of variables by a much smaller set of derived variables which still retain most of the relevant information. This process of condensing a set of variables into a few new variables is usually called ordination and the new variables are often referred to as ordination axes. Ordination tries to approximate the complex pattern of the full data set in a few dimensions. If the reduction in dimensionality is sufficient the results can be presented visually by plotting graphs of the new variables against each other. Thus ordination allows us to visualize the relationships between sites, samples, species, variables, etc. much more easily. It may be used as a technique in its own right or as a preliminary to other forms of analysis based on the derived variables.

6.3.1 PRINCIPAL COMPONENTS ANALYSIS (PCA)

Principal component analysis (PCA) was the first ordination technique to be developed and is probably still the most used method of ordination today. In PCA linear combinations of the original variables are created that embody as much as possible of the variance in the data. The first new variable, or principal component axis (PC), is chosen to account for the maximum amount of variance possible in a single variable and subsequent PC axes are chosen to explain as much as possible of the remaining variance while being uncorrelated with previously derived axes. New components can be calculated until all of the original variance is accounted for and the maximum number of possible components is the same as the number of original variables. The complete set of PCs therefore forms a set of new variables which embody successively smaller proportions of the original total variability. The rationale behind PCA is that if the first few components account for most of the variance, then hopefully they also represent most of the important information in the data and the remainder can be ignored, thus reducing the number of variables to be analysed.

Principal components are usually derived from either the raw data (via the covariance matrix) or from standardized data (via the correlation matrix) though other preliminary transformations of the raw data can be used. Using standardized data can be thought of as giving all variables equal importance regardless of their scale of measurement (see section 6.2.1). Whether this is sensible depends on the individual situation being examined. Users should be aware of which method their statistical software uses by default. Standardization is appropriate when variables involve a range of scales and types of measurement (e.g. concentration in milligrams per litre, weight in grams, and plant stem height in centimetres). Using unstandardized data in PCA gives greater emphasis to variables with the greatest variance. For instance in vegetation surveys where the variables are the abundance or cover of individual species, using unstandardized data will place less emphasis on rare or uniformly common species and greater emphasis on those species whose abundance or cover is most variable; the first few axes may be dominated by just one locally very abundant species. In the same situation standardization would give equal emphasis to all species, including the uncommon ones. An alternative to standardization is to transform the data matrix before analysis so as to reduce the disparity in variation between variables. In the case of data on chemical concentrations or species abundances a log transformation is often effective, while a square root transformation may be helpful for percentage ground cover.

PCA will only perform effectively as an ordination technique if the first few PCs (ideally two) account for a substantial proportion of the total variance in the original data. If the original variables are uncorrelated then the components derived from PCA turn out to be just the original variables arranged in order of variance. In this situation there will be no reduction in dimension and hence no point in using PCA. As with all ordination methods it is common to plot the first few axes against one another and make a subjective assessment of the resulting patterns. PCs can also be used in subsequent forms of analysis in place of the original variables (an example is principal components regression as used by Mullis *et al.* 1996).

Examination of the coefficients of individual PCs may give an idea of what the PC represents. Variables with large coefficients contribute most to a PC and the signs of the coefficients show which variables have similar effects. Interpretation of coefficients is simplest when using standardized data where the coefficients represent only the contribution of its associated variable. For

Using standardized data

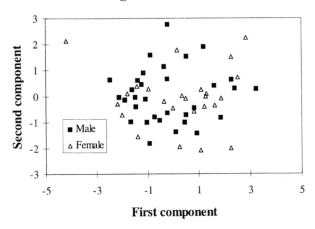

Using unstandardized data

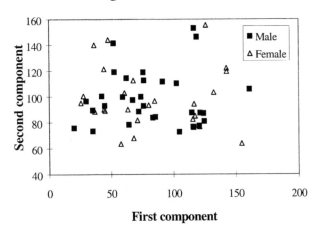

Figure 6.2 Principal component analysis of starling data (32 males and 25 females) using both standardized and unstandardized data.

unstandardized data the coefficients are also affected by the variances of the variables.

Figure 6.2 and Table 6.1 show the results of the ordinations of the starlings data set (Appendix 3) based on the six physical variables (omitting load variables) for both standardized and unstandardized data. Using standardized data (i.e. the correlation matrix) the first two axes account for 42% and 20% of the total variance and the first four components are needed to account for 91%. With the possible exception of protein concentration, the coefficients for the

Table 6.1 Principal component analysis of starling data for six physical variables

Using the correlation matrix

	PC1	PC2	PC3	PC4	PC5	PC6
Percentage of variance	42	20	18	12	6	4
Cumulative percentage	42	61	79	91	96	100

	PC1	PC2	PC3	PC4	PC5	PC6
Liver weight	0.40	−0.59	−0.08	0.20	0.67	−0.01
Lipid concentration in liver	0.38	−0.28	−0.41	−0.68	−0.33	−0.21
Protein concentration in liver	0.19	0.69	−0.49	−0.12	0.48	0.03
Body lean weight	0.50	0.14	0.45	−0.25	−0.04	0.68
Body lipid weight	0.43	0.03	−0.40	0.65	−0.47	0.15
Weight of pectoral muscle	0.48	0.27	0.47	0.10	−0.04	−0.68

Using the covariance matrix

	PC1	PC2	PC3	PC4	PC5	PC6
Percentage of variance	73	26	1	0	0	0
Cumulative percentage	73	99	100	100	100	100

Variable	PC1	PC2	PC3	PC4	PC5	PC6
Liver weight	0.01	−0.01	0.06	−0.05	1.00	−0.03
Lipid concentration in liver	0.99	−0.16	−0.04	0.01	−0.01	0.00
Protein concentration in liver	0.16	0.99	−0.06	0.00	0.01	−0.00
Body lean weight	0.02	0.00	0.08	−0.98	−0.06	−0.15
Body lipid weight	0.04	0.05	0.99	0.09	−0.05	−0.01
Weight of pectoral muscle	0.00	0.00	0.03	−0.15	0.02	0.99

first component give roughly equal weight to each variable. Thus the first axis is
an average of the six measurements and might be described as a 'size' variable.
The second component gives most emphasis to liver weight and protein concentration but with opposite signs. Starlings that score highly on this
component will have small livers and high protein concentrations.

Using the unstandardized data (i.e. the covariance matrix) the differences in
the proportion of variance accounted for by each component are much greater
with the first two axes representing 73% and 26%. However, each component is
dominated by a single variable, lipid concentration and protein concentration
respectively for the first two components. Because the six original variables
have very different levels of variability the PCA has simply placed the variables in

order of variance and therefore adds little or nothing to our understanding of this data set. This demonstrates the difference between PCA ordinations based on the two matrix types.

Additional information can be incorporated in ordination diagrams in a variety of ways. Qualitative characteristics, for example, can be included by varying the plotting symbol. In Figure 6.2 male and female birds are indicated by different symbols. There is no clear difference in the position or spread of the two sets of points, indicating that for this data set the main sources of variability are not sex related.

It is common for PCA and other ordinations to be represented by biplots, defined as the joint representation of the rows and columns of a data matrix. The usual practice is for rows (sites, individual birds, etc.) to be represented as points and columns or variables (e.g. metals) to be represented by vectors. The lengths of the vectors indicate the variability associated with the variable and the cosine of the angle between vectors reflects the correlation between them. Characteristics of individual points can be determined from their position relative to vectors (see Digby and Kempton 1987 for more details).

Figure 6.3 is a biplot representation of the starling metal data. The vectors representing the metals all lie to the right of the origin indicating that the first

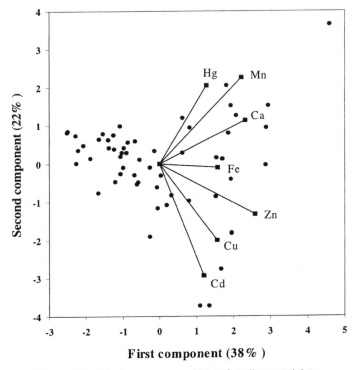

Figure 6.3 Principal component biplot of starling metal data.

axis is again a size variable which can be interpreted as a measure of overall metal load. The second axis contrast of Hg, Mn and Ca with Cd, Cu and Zn is also clear. The points represent individual birds. Those that lie to the right of the origin and hence closer to the metal vectors will be those with the highest liver metal concentrations while those to the left of the plot will have lower concentrations. Note that points on the left, those with low loadings, are clustered together, indicating that they tend to have low values for all metals. In contrast, points on the right-hand side are more spread out, indicating that birds with higher metal concentrations are more variable in their metal concentrations. A clear outlier can be seen in the top right-hand corner of the figure, in the direction of the manganese vector. This point represents the starling with the very highest manganese concentration. In contrast the three birds at the bottom of the figure have very high cadmium concentrations but only moderate levels of manganese.

In recent years a number of extensions or modifications of PCA have been developed. Common principal components (Krzanowski 1988) for example is a technique used for the simultaneous analysis of several data sets which constrains the PCs derived from each dataset to have the same coefficients, thus facilitating comparison of results across datasets. Projection pursuit (Krzanowski 1988) is a technique which maximizes not the variance of components but some other quantity. A measure of 'clumpiness' could be used, for example, if the main concern was to find groups in the data.

6.3.2 CORRESPONDENCE ANALYSIS

Principal component analysis was developed as an ordination technique for use with quantitative variables, representing measurements along a numerical scale, though it has also been used for ordinal data, where values represent an order rather than an exact measurement. For qualitative measurements such as leaf shape, however, PCA is not appropriate. Qualitative information can be coded numerically, for example with leaf shape, round = 1, oval = 2 and pointed = 3, but using PCA on such categorized data is equivalent to assuming that the values, and their differences, are quantitatively meaningful rather than just a convenient coding system. The key difference between PCA and correspondence analysis (CA) is that CA was derived for us with categorical data while PCA is primarily intended for continuous data. Mathematically, however, the two techniques are very similar.

To use CA each categorical variable is transformed to a set of binary variables, one for each category. For example, the single variable leaf shape, coded 1 to 3, would be replaced by three variables, 'round', 'oval' and 'pointed', each coded 0 or 1. Thus the whole dataset is reduced to a matrix of zeros and ones. CA can be thought of as applying a weighted form of PCA to this matrix. In practice, most CA software will accept the original categorical variables and perform its own transformation to obtain the data in an appropriate form. As with PCA the resulting ordination is frequently presented as a biplot showing the first two ordination axes (or a set of biplots if more than two ordination axes are retained). Each biplot will contain a set of points representing the units and a set of points representing the categories of the variables. Note that, whereas PCA produces one point (or vector) for each original variable, CA produces several.

Although it is not appropriate to use PCA for datasets containing qualitative variables, CA can be used for any dataset since continuous variables can always be categorized. With chemical concentration data, for example, cut-off levels can be used to categorize variable values as high and low or high, medium and low. For such datasets a key element in the choice between PCA and correspondence analysis is that PCA can reconstruct environmental gradients to which the variables in the dataset have a linear response while, under certain specific models, CA and its associated techniques can reconstruct an environmental gradient to which the response is unimodal, i.e. there is a point along the gradient for each variable (e.g. species) at which a maximum response is achieved. For species the point is the centre of its niche and is where the species is either most abundant or, for presence/absence data, where it is most likely to be found. The argument for CA is that there are many environmental stimuli for which a maximum can be expected. In the case of plants, greatest abundance might be expected within a range of wetness values; lower abundance occurring at lower (drought) and higher (waterlogged) values. How well this concept of unimodality equates to ecotoxicology work remains to be seen and is critically dependent on the nature of the study under investigation. In the case of mortality or sub-lethal effects one might expect a monotonic relationship to increasing levels of environmental pollution and PCA might be considered to be more appropriate. On the other hand a tolerant species might increase in abundance as moderate levels of a chemical pollutant eliminate its less tolerant predator species, but eventually succumb at higher concentrations. This phenomena occurs for freshwater macro-invertebrates subject to progressive river eutrophication. In general, different ordination methods will produce slightly different results from the same dataset. Which method is most appropriate will depend on the data and the purpose of the analysis. Each study situation is unique and should be considered on its own merit.

An example of CA is given in Figure 6.4. This again shows the starling metal data and is directly comparable with Figure 6.3. To obtain Figure 6.4 the values for each metal were ranked and divided into three, roughly equal-sized, groups of high, medium and low values. The categorized data was then analysed using CA. Not surprisingly the overall pattern produced by CA is similar to that produced by PCA; high categories for all metals occur on the right-hand side, low values on the left and medium values closer to the origin. Thus the first component is again a measure of overall loading. The second component is also similar to that found by PCA; high levels of Hg occur at the top right of the

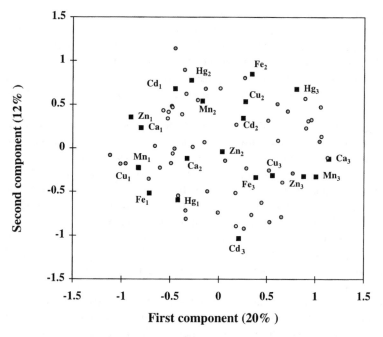

Figure 6.4 Correspondence analysis biplot of starling metal data. Values for each metal categorized as low (1), medium (2) or high (3).

graph and high levels of Cd at the bottom right. There are, however, also some differences. The points are more evenly spread than in Figure 6.3 with no real outliers. This is a result of the categorization which reduces the influence of extreme values. The point representing high Mn is much lower than in the PCA diagram suggesting that its PCA position may have been influenced by the outlier discussed above.

A criticism of CA, and indeed also of PCA, is that the second axis, which is mathematically derived to be linearly uncorrelated with the first, may in fact be a quadratic variant of it. This manifests itself by the plotted points forming an arch or horseshoe shape when the first two axes are plotted against one another. This can happen with sites by species data when sites at one end of a long gradient have few or no species in common with sites at the other end. One way to avoid this is to use the first and third axes. Another is to detrend the second and subsequent ordination axes as they are derived to ensure that they have no relationship with previous axes. Such a modification is known as detrended correspondence analysis (DCA) and has been widely used in ecology, primarily as a result of the availability of the DECORANA (Hill 1979) program subsequently incorporated within the CANOCO package (Ter Braak 1988).

6.3.3 FACTOR ANALYSIS

The aims of factor analysis (FA) are similar to those of PCA, to replace a set of observed variables by a smaller set of derived variables. In FA the observed variables are assumed to be linear combinations of a set of unobserved variables, called factors, plus a random error. This is in contrast to PCA where the new variables, or components, are derived directly as linear combinations of the observed variables. This rather subtle distinction leads to different estimation techniques, but in practice the results of applying the two methods to the same data set are often very similar as are methods used to interpret factor coefficients (called loadings). Factor analysis was developed in psychology at the turn of the century as a means of explaining the associations between test results in terms of a small number of distinct abilities such as general intelligence, numerical ability, verbal ability, etc.

PCA and FA are often confused in the literature and this confusion is made worse by one popular factor estimation procedure being known as principal factoring. In addition, some packages use principal components as starting values for factor analysis. Unlike PCA, FA is unaffected by standardizing or rescaling the observed variables. The most important difference between factor analysis and PCA, however, is that FA is a parametric technique with a distributional model for the data. As a consequence formal statistical tests can be applied to FA results. While being very popular in the social sciences, FA has, unjustifiably, a poor image in many circles. Chatfield and Collins (1980) provide a detailed discussion of the supposed drawbacks to FA while a less negative discussion can be found in Krzanowski (1988).

Factor analysis is just one of a number of techniques which aim to explain the associations or distribution of an observed set of data in terms of a set of unobserved or latent variables. Factor analysis assumes that the observed and latent variables are all continuous. In fact the assumption is that together they have a multivariate normal distribution. Latent-structure analysis assumes that the observed variables are discrete and the latent variables continuous. Latent-class analysis assumes both observed and latent variables are discrete, and mixture modelling assumes continuous observed variables and discrete latent variables. Collectively these techniques are referred to as latent variable models (Krzanowski 1988, Chapter 16).

6.3.4 MULTIDIMENSIONAL SCALING / PRINCIPAL COORDINATE ANALYSIS

Multidimensional scaling (MDS) is an ordination technique which derives new axes to give the 'best' agreement of the relative distances between pairs of units in the ordination space and their actual pairwise dissimilarities. The great flexibility of MDS compared to the other ordination techniques mentioned above is that dissimilarity (or similarity) between pairs of units may be calculated by the user in a wide variety of different ways, as discussed in section 6.2.2. The technique is particularly useful when the original data contain some missing data and more conventional ordination packages cannot be used. The aim is usually to derive a diagram showing the best two-dimensional representation of similarity between the units.

There are two forms of multidimensional scaling, metric and non-metric, and the metric form is known by a variety of other names including principle coordinate analysis (PCO). The non-metric form tries to maximize agreement between the ranks of the pairwise dissimilarities and the ranks of distances in the ordination diagram instead of the actual distance and dissimilarity values. One advantage of non-metric MDS is that because ranks rather than actual distances are used, a few unusual outliers are not allowed to completely dominate the ordination. Multidimensional scaling is popular within sectors of the marine ecological sciences, where the non-metric form of MDS is used (Clarke and Ainsworth 1993). If Euclidean distance between units based on the original variables (either unstandardized or standardized) is used as the dissimilarity measure then MDS is equivalent to PCA (based on the covariance and correlation matrices respectively) and will provide effectively the same ordination diagram.

6.4   ANALYSING MULTIVARIATE GROUPED DATA

In many situations the units of the data set fall, by chance or design, into distinct groups and the main interest of the researcher is in comparing the groups in terms of the recorded attributes. Examples include designed experiments whereby one or more groups of individuals are subject to different treatments or impacts, while another group is used as the control; comparing males and females for toxicity levels; or assessing environmental differences between taxonomic groups. In other research, the primary aim is not to compare the groups but to derive equations or quantitative rules, based on the available variables, which can be used to discriminate between the groups and perhaps to allocate new individuals, whose group is unknown, to one of these groups. A third objective may be to visually depict the differences between groups. These three different objectives lead to different multivariate techniques, multivariate analysis of variance (MANOVA), discriminant analysis (DA) and canonical variates analysis (CVA) respectively.

Although the mathematical principles and assumptions underlying these techniques are the same irrespective of the number of groups or treatments involved, it is useful to first consider the simplest situation with just two groups and then extend this to several groups.

6.4.1 HOTELLING'S T^2

When the observations fall into two groups (e.g. samples from a control and treatment site), then for a single quantitative response variable, it is common to test for a difference in mean between the two groups using a Student's t-test. Hotelling's two-sample T^2 is the multivariate equivalent of Student's t-test. It is used to assess whether there is a difference between two groups when a number of variables are considered together.

As an example we have used the seven metal determinations from the starling data to investigate whether a difference exists between male and female starlings (Table 6.2). Despite a significant univariate result for Fe ($p = 0.005$), the overall T^2 test statistic is not significant at the traditionally defined level of $p = 0.05$, although the calculated level of $p = 0.09$ is low enough to cause us not to completely dismiss the possibility of a difference. This highlights one advantage of using multivariate tests. With T^2 the probability of a type 1 error remains constant, whereas the use of multiple univariate t-tests increases the chance of reaching an incorrect decision. For instance in seven independent tests the probability of obtaining a significant ($p < 0.05$) result when no real difference exists would be approximately 30% ($= 1 - 0.95^7$). The significant Fe result is therefore not surprising.

In other situations it is quite possible to obtain significance in T^2 when there is no significant difference for individual variables because T^2 uses the joint variability of all the variables (i.e. their variances and inter-correlations) and hence is able to detect differences between groups which are either not present

Table 6.2 A comparison of the mean log transformed concentrations $\pm$ standard deviation s.d. (untransformed mean given in parentheses) of ($q = 7$) metals in livers of ($n_1 = 32$) male and ($n_2 = 25$) female starlings using two-sample, two-sided Student t-tests and Hotelling multivariate T^2 test; $p =$ test significance probability. (Under the null hypothesis of no differences in means between groups:

$$T^2 (n_1 + n_2 - q - 1)/(q(n_1 + n_2 - 2))$$

has an F-distribution with q and ($n_1 + n_2 - q - 1$) degrees of freedom)

Metal	Male mean $\pm$ s.d.	Female mean $\pm$ s.d.	Two-sample t-test	
			t	p
Zn	4.33 ± 0.10 (77.3)	4.29 ± 0.20 (74.6)	0.74	0.46
Cu	2.86 ± 0.18 (17.8)	2.85 ± 0.29 (18.1)	0.12	0.91
Fe	7.93 ± 0.40 (2981)	7.55 ± 0.59 (2198)	2.91	0.005
Ca	5.29 ± 0.53 (230)	5.33 ± 0.46 (232)	0.32	0.75
Mn	1.57 ± 0.59 (5.51)	1.43 ± 0.32 (4.38)	1.13	0.26
Cd	1.28 ± 0.41 (3.92)	1.21 ± 0.60 (3.96)	0.57	0.57
Hg	-1.20 ± 1.46 (1.01)	-1.57 ± 1.53 (0.98)	0.90	0.37
			$T^2 = 14.92$	0.09

or not large enough to produce significance for any individual variable (see Figure 6.1).

6.4.2 MULTIVARIATE ANALYSIS OF VARIANCE (MANOVA)

In much the same way that ANOVA is the extension of the Student's t-test to more than two groups or treatments when response variables are analysed individually, multivariate analysis of variance tests (MANOVA) are the extension of Hotelling's T^2 to several groups (and also the multivariate equivalent of ANOVA).

There is some contention about the appropriate test statistic to use for significance testing in MANOVA. In the example below we have used Wilks' lambda (Λ) which is the likelihood ratio statistic. Like T^2, MANOVA is less resistant to departures from multivariate normality than ANOVA. For a discussion of alternative test statistics see Chatfield and Collins (1980) or Krzanowski (1988).

As an illustration of MANOVA we use the metals in the starling data to see if there exist seasonal differences in metal burdens of the birds. A summary of the analysis is shown in Table 6.3. Significance ($p < 0.05$) was achieved in four of the univariate F-tests and overall from the MANOVA, indicating that there were highly significant seasonal differences in metal burdens of starling livers. Spring was associated with high Cu and low Fe levels, while winter was associated with low Cd and high Hg. The latter result may be in consequence of the intake of Hg from seed dressings while foraging in arable fields.

Table 6.3 Multivariate assessment (MANOVA) of seasonal differences in the mean log transformed metal concentrations ($\pm$ standard deviation) of captured starlings based on Wilks' lambda (Λ). Univariate ANOVA F-tests are also given, n = sample size per season. (Under the null hypothesis of no differences in means between groups:

$$(N - g - (q - b + 1)/2) \log_e \Lambda$$

has approximately a chi-square distribution with qb degrees of freedom, where b is the minimum of q and $(g - 1)$, q = number of variables = 7, g = number of groups = 4)

Metal	Winter $n = 17$	Spring $n = 10$	Summer $n = 15$	Autumn $n = 15$	Two-sample t	t-test p
Zn	4.37 ± 0.17	4.37 ± 0.27	4.26 ± 0.15	4.33 ± 0.21	0.61	0.62
Cu	2.83 ± 0.17	3.04 ± 0.26	2.73 ± 0.25	2.89 ± 0.18	4.69	0.006
Fe	7.80 ± 0.34	7.27 ± 0.71	7.70 ± 0.48	8.11 ± 0.30	6.93	0.001
Ca	5.47 ± 0.50	5.27 ± 0.55	5.14 ± 0.50	5.32 ± 0.44	1.23	0.31
Mn	1.70 ± 0.47	1.37 ± 0.23	1.45 ± 0.30	1.44 ± 0.72	1.35	0.27
Cd	0.85 ± 0.43	1.25 ± 0.52	1.47 ± 0.35	1.49 ± 0.42	8.00	<0.001
Hg	-0.37 ± 1.66	-1.64 ± 0.86	-2.29 ± 0.64	-1.37 ± 1.66	5.65	0.002
					$\Lambda = 14.92$	<0.001

It should be remembered that tests based on two-group T^2 or the more general MANOVA assume that the within-group distributions of variables are multivariate normal with the same variances and covariances in each group. If the within-group standard deviation for a variable tends to increase with the group mean, then, as in univariate analyses, a square root or log transformation of that variable can make the variances more equitable and the distribution less skewed (e.g. as for the starling data in Table 6.2).

6.4.3 TWO-GROUP DISCRIMINATION

In certain circumstances the aim of analysis is to develop rules based on the available variables for classifying units into externally defined groups. It is assumed that a dataset, known as the training set, is available for which group membership of the units is known. This is used to derive the required rules which are then used to classify other or future units for which group membership is not known. Often, as a test of the effectiveness of the method, the derived rules are used to classify the training set. This gives an over-optimistic result, especially if sample sizes are small and/or the number of variables is relatively large, since the derived rules will be optimum for the data from which they were derived. A more realistic estimate of the discrimination potential can be obtained by jack-knifing or cross-validating results, whereby each individual in turn in the training set is temporarily excluded and then allocated to a group based on a rule developed from all the other individuals. The discrimination potential is measured by the proportion of individuals allocated to their true group.

A wide variety of discrimination techniques are available (see Seber, Chapter 6, 1984). However, by far the most commonly used is linear discriminant analysis (LDA) which is based on the same assumptions as MANOVA, namely multivariate normality within each group with equal variances and covariances of the variables within each group. Using this model, linear discriminant functions (LDFs) are derived which are combinations of the original variables that maximally discriminate between the groups. These LDFs are calculated in various alternative forms in different software packages, but they are all mathematically equivalent and give the same results when used to allocate individuals to groups.

The use of two-group discrimination to classify individuals is shown by data used to sex oyster-catchers from their morphology as part of studies to assess differences between the sexes in their feeding habits (Durell *et al.* 1993) and their susceptibility to heavy metal loadings in their estuary. Oyster-catchers have no external characteristics that readily permit sexual identification, even in the hand. However, after trial T^2-tests with various subsets of the body measurement variables, the following LDF based on just two variables, bill depth (mm) and head plus bill length (mm), was found to discriminate the sexes well (Figure 6.5):

$$\text{LDF} = S = 0.187 + 6.152 \text{ bill depth} - 0.555 \text{ (head + bill length)} \quad (6.1)$$

The difference between the two group means on the LDF is referred to as the Mahalanobis D^2 distance. On the assumption of multivariate normality and equal variability within groups, S is equal to the log of the ratio of the likelihood that the observation comes from group 1 (in this case, males) to the likelihood that the observation comes from group 2 (females). The probability of any individual being male can therefore be estimated from its discriminant score S by

$$Pr(male) = 1/(1 + e^{-S}) \qquad (6.2)$$

as shown for the oyster-catcher data in Figure 6.5. Individuals in most forms of multivariate discrimination are allocated to the most likely group. In terms of S, this means classifying an oyster-catcher as male if S is greater than zero. Using the above LDF, all except one of the 40 male oyster-catchers would be correctly sexed and only 3 of the 38 females would have been misclassified as males, giving an overall 95% of birds correctly classified (for follow-up analyses involving this method of sex determination, it may be sensible to only use those birds that could be classified with at least 90% probability (e.g. all except 6 male and 12 female oyster-catchers (Figure 6.5)).

6.4.4 MULTIVARIATE DISCRIMINATION BETWEEN SEVERAL GROUPS

When there are more than two groups LDA techniques produce one discriminant function for each group. Effectively these functions can be thought of as being proportional to the likelihood of an observation belonging to the associated group. Units whose group membership is unknown are allocated to the group for which the value of the associated discriminant function is greatest, i.e. to the most likely group.

For illustration only, we demonstrate the general technique here by using LDA with the starlings' metals data to assess the extent to which the metal loadings within a starling identify its season of capture. The jack-knifed estimates indicate that 53% of birds could be correctly allocated to season on the basis of their liver metal concentrations (Table 6.4). Since these seasons are, in effect, a continuum artificially separated we should not be surprised that discrimination is not better achieved. It should be remembered that if allocation of birds to seasons were completely at random (i.e. guessing) then we would only expect 25% correct allocation.

If the assumption of equal variances and covariances is not reasonable they can be allowed to vary from group to group. This produces discriminant functions which are quadratic combinations of the original variables. However, quadratic discrimination is very sensitive to departures from normality, especially skewness, and can only be used effectively with large samples. Hence it is generally not appropriate. It is better to try to transform the variables to give roughly equal within-group normal variability.

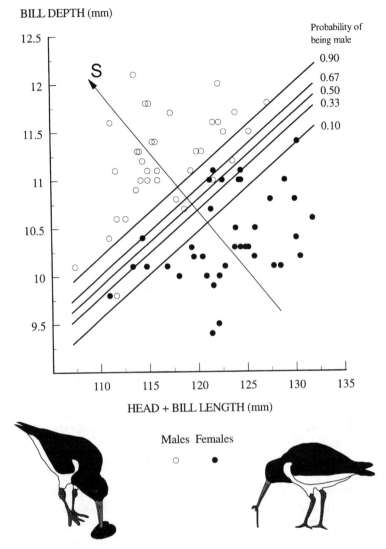

Figure 6.5 Sexing oyster-catchers by morphometric discrimination. *S* denotes the line of the linear discriminant function (LDF) (equation 6.1) based on 40 males (o) and 38 females (●). Straight-line contours of the probability of an oyster-catcher being male are given.

6.4.5 CANONICAL VARIATES ANALYSIS (CVA)

When analysing grouped data it is often useful to visually depict group differences, either for subjective interpretation or as a visual aid to MANOVA and discriminant analysis. This is most commonly done by creating new ordination axes (see section 6.2 and Box 6.2) from the original variables which separate the

Table 6.4 Linear discriminant analysis of log transformed starling metal data, showing allocation of individual birds to groups (seasons) and the same allocation under jack-knifing

		Normal allocation true group				Jack-knifing true group			
		W	Sp	Su	A	W	Sp	Su	A
Allocated	W	**12**	0	1	3	**10**	2	1	3
	Sp	0	**7**	2	0	1	**5**	2	1
	Su	4	1	**9**	4	4	1	**8**	4
	A	1	2	3	**8**	2	2	4	**7**
Total number		17	10	15	15	17	10	15	15
% Correct		71	70	60	53	59	50	53	47
%Correct overall		63				53			

known groups as much as possible. Interpretation of the relative importance of the original variables to these axes can then help the user understand the ways in which the groups differ from one another. These new axes are called canonical variates and the technique used to derive them is known as canonical variates analysis (CVA). Each successive axis is chosen to maximally separate the groups while being uncorrelated within groups with previous axes. In mathematical terms the axes are chosen to maximize the ratio of the between-group to within-group variance. The maximum number of canonical variate axes is limited to the smaller of the number of variables or one less than the number of groups. Thus, with only two groups there is only one canonical variate, which is equivalent to the LDF of section 6.4.3. For more than two groups, however, there is no simple relationship between canonical variates and discriminant functions.

In presenting the results of CVA it is usual to plot the group means or centroids using the first two canonical variates. Each canonical variate is usually scaled for convenience so that the overall mean for the training set individuals is zero and the average (weighted by group sample size) within-group standard deviation on the axis is one. If the number of units is small enough they can also be plotted to give an idea of the variability in the data. Alternatively it is possible to create confidence ellipses for either the group centroids or the individuals within each group, but the latter especially relies on having multivariate normality whereas CVA, unlike MANOVA or LDA, is not based on this assumption.

The same starling metals data used to illustrate LDA have been used to show the separation achieved by CVA (Figure 6.6). The separation achieved is not very good, although groups of starlings measured in the spring and summer tend to have distinctly different metal loadings (see section 6.4.2 and Table 6.3). There is more overlap in this example than might be expected in situations

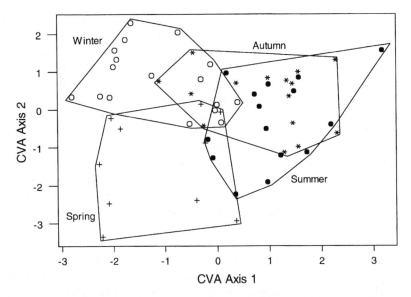

Figure 6.6 Canonical variates plot of log transformed starling metal data grouped by season (see Table 6.3 for further data details).

where groups were formed from more discrete categories, for example in treated versus untreated comparisons.

6.5 CLUSTER ANALYSIS

The purpose of cluster analysis is to identify groups or clusters of similar sites or species within a data set, although clustering techniques are sometimes used when there is little evidence of, or reason to expect, distinct groups. In such cases the clustering procedure can be thought of as simply a convenient way to summarize the enormous variation in a large number of units by grouping 'similar' units together. For example, Wright *et al.* (1984) classified a large range of unstressed river sites in Great Britain into 25 groups on the basis of their macro-invertebrate composition to provide a simple method of summarizing biological types of site. (Incidentally, they then used LDA equations (section 6.4) based on the physical and chemical conditions at each type of site to predict the expected fauna in the absence of any pollution.) It should always be remembered, however, that when used in this way the groups obtained from cluster analysis have no objective reality and will not usually be stable across different sample datasets. In contrast when used as intended for data containing 'real' groups, the groups identified by cluster analysis should be stable and have an objective existence for which later evidence may be found.

There is a bewildering variety of methods for clustering individual units.

These methods can be broadly divided into hierarchical or non-hierarchical techniques, where hierarchical means that individual units are put into a hierarchy of groups with a tree-like branching structure represented by a dendrogram (e.g. groups A and B at one level of the hierarchy are both part of super-group H, say, at a higher level). Hierarchical techniques are themselves subdivided into agglomerative and divisive methods. Agglomerative techniques start with each unit as a separate cluster and gradually combine individual units and groups of units until a single group or small number of groups remain. This can be considered a bottom-up approach. The alternative divisive hierarchical approach might be considered a top-down approach as, starting with all the units in the same group, the set of units is gradually and successively split into more and more clusters. Divisive hierarchical clustering techniques are little used in practice, largely because of the computational resources needed to form an optimal division of a large group of units.

The most frequently used non-hierarchical method of clustering is K-means clustering. Here the number of groups is chosen beforehand and, starting from an initial 'seed' unit in each group, the remaining units are allocated to the group to which they are most similar, where dissimilarity is measured by squared Euclidean distance of the unit from the group's centroid or mean. After all units have been allocated they may be swapped between groups in an attempt to improve the separation between the groups as obtained by minimizing some criteria such as the sum of within-group squared distances.

Agglomerative hierarchical techniques are the most commonly used cluster methods. All agglomerative hierarchical methods start with a matrix giving the similarity or distance between each pair of units. This can be calculated in a number of ways; the most appropriate should be dictated by the form of the data, not the software package most readily available (see section 6.2.2 and Box 6.1). At any stage in an agglomerative clustering approach, the next two units/groups joined are those with the least 'inter-group distance' (or highest 'inter-group similarity'). The definition of inter-group, as opposed to inter-unit, distance depends on the linkage method used. Single linkage, for example, uses the minimum distance between a unit in one group and a unit of the other group and is thus also referred to as the 'nearest neighbour' method, whereas complete linkage or 'furthest neighbour' uses the maximum such distance. The common linkage methods are listed in Box 6.3.

The combined choice of a distance measure and of a linkage method results in a bewildering set of possible options of which there is no universally agreed optimum. It therefore seems sensible to try several methods; if clustering is robust then one might expect broadly similar clustering from different approaches although different linkage methods are appropriate for different situations. Single linkage, for example, is prone to 'chaining' of units leading to long, thin groups, while complete linkage tends to produce roughly equal-sized groups. The results of any hierarchical clustering is usually represented visually in the form of dendrograms (tree diagrams).

Box 6.3   Commonly used measures of distance between two groups of units

(a) Nearest neighbour (single linkage)
Distance between two groups is the shortest distance between any point in the first group and any point in the second group.

(b) Furthest neighbour (complete linkage)
Distance between two groups is the furthest distance between any point in the first group and any point in the second group.

(c) Group average
Distance between two groups is the average of all possible distances between any point in the first group and any point in the second group.

(d) Median
Distance between two groups is the median of all possible distances between any point in the first group and any point in the second group.

(e) Centroid
Distance between two groups is the distance between the means of the two groups.

(f) Minimum variance (Ward's method)
Joins the two groups for which the increase in overall within cluster variance is least.

As an example, we might consider how the starlings can be grouped according to the concentrations of seven metals in their livers. To simplify presentation, a random subset of 20 starlings was used (Figure 6.7). When based on single linkage using Euclidean distance on the log transformed metal variables, there is a tendency for most birds slowly to join into one large group, but with one triplet of birds (1, 5 and 6) maintaining a completely separate group and two birds (10 and 13) appearing to be different from the rest and each other as linkage to other clusters is very delayed. Figure 6.7b shows the same data, but clustered with average linkage; many of the basic features are similar to those of single linkage, although bird 13 is no longer classed as different. The above two cluster analyses use Euclidean distance on log trans-formed concentrations, which gives more weight in measuring distance to those metals which vary by the highest order of magnitude (e.g. the standard deviation of log Hg is 1.50 compared to only 0.19 for Zn). To force all metals to contribute equally to the between-bird distance measure, the clustering was repeated using average linkage for Euclidean distance on the standardized metal concentrations (Figure 6.7c). This retains the triplet group (1, 5, 6), gives no apparent outliers, but shows reasonable similarity with the other methods. Interestingly, using the non-hierarchical *k*-means clustering to produce four

(a)

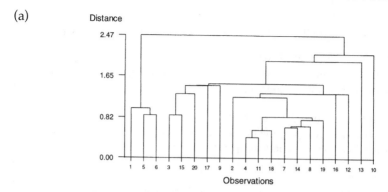

(b)

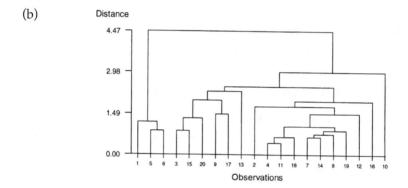

(c)

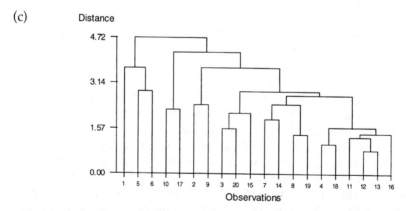

Figure 6.7 Three cluster analysis dendrograms for 20 randomly selected birds from the starling dataset using the seven metal concentration in liver variables. The clusterings were based on (a) Euclidean distance on log transformed values using single linkage, (b) as (a) but using average linkage, (c) Euclidean distance based on standardized values of the original concentrations to give each variable equal importance.

groups from the log-transformed concentrations gives exactly the same grouping as average linkage above when the dendrogram (Figure 6.7b) is cut at the four group level, namely groups (1, 5, 6), (3, 15, 20, 9, 17, 13), (2, 4, 11, 18, 7, 14, 8, 19, 12, 16) and (10).

6.6 EXAMINING RELATIONSHIPS BETWEEN TWO SETS OF VARIABLES

In many situations the interest of the researcher may not be with groups of units but with groups of variables. In ecology, information is often collected on both species composition and environmental factors with the aim of relating one set of variables to the other. The carabid data in Appendix 2 is an example of this. Many forms of analysis have been devised for this situation. Most are ordination techniques where new axes are formed from one or both sets of variables, but in which the axes are constrained to represent the relationships between the two sets of variables. For this reason ordination, sometimes referred to as gradient analysis, techniques are often referred to as being of two basic forms, indirect and direct. The former, described above, start with a single set of variables, for example, a site by species data matrix, and axes derived from this may be interpretable as gradients of a particular type, e.g. environmental, but without any direct data input from this area of interpretation. Direct gradient analysis on the other hand starts with two sets of variables, typically species data and environmental data for a number of sites. The two sets of variables are represented on the same ordination diagram and assessment of relationships between sites, species and environmental variables made directly from this. Direct ordination methods are also known as canonical ordination methods.

The techniques described in this section are mathematically more complex than the techniques described so far. Only a brief overview is therefore given here. Readers should refer to more extensive texts for full details (Jongman *et al.* 1995; Ter Braak 1988).

6.6.1 MULTIVARIATE MULTIPLE REGRESSION

We start with a technique that is not a form of ordination. Multivariate multiple regression is a direct extension of multiple regression (see Chapter 5) in which we are concerned with several response variables instead of just one. The end result of the analysis is a set of regression equations, one for each response variable, but these equations are derived to take account of the inter-relationships between the response variables as well as their relationship to the explanatory variables.

6.6.2 CANONICAL CORRELATION ANALYSIS

Canonical correlation analysis can be viewed as a canonical form of PCA. The variables measured on each unit are divided into two sets and a linear combination of the first set of variables is created that maximizes correlation with a linear combination of the second set (see Box 6.2). These might be, for instance, species variables and environmental variables. As with most ordination techniques more than one pair of axes can be derived and successive pairs of axes are chosen to maximize the correlation in the remaining unexplained variation, but subject to the restriction that each axis must be uncorrelated with all previous axes. Thus canonical correlation analysis can be thought of as extracting the correlation between the two sets of variables and concentrating it in a few new pairs of variables, in contrast to PCA which takes the variability of a single set of variables and concentrates it into a few new axes. For illustration we turn again to our starling data and compare the physical and metal measurements. Table 6.5 shows the first two pairs of axes and the corresponding correlations (the canonical correlations) between the axes in each pair. Taking the first pair the physical measurements axis can be interpreted as liver weight minus body weight, while the metals axis has a particularly high coefficient for Cd. The correlation between these two axes is -0.79. This suggests that birds with relatively large livers for their overall body size had relatively low levels of Cd (and some other metals) in their livers.

Table 6.5 Canonical correlation analysis for starling data. Correlations derived between six physical measurements and seven log transformed metal determinations

	Axis1	Axis 2
Canonical correlations	−0.79	0.58
Physical measurements		
Liver weight	0.689	−0.378
Lipid concentration	0.243	0.197
Protein concentration	0.304	−0.586
Body lean weight	−0.488	1.145
Body lipid	0.353	0.619
Pectoral muscle weight	0.008	−0.693
Metal concentrations		
Zn	0.213	−0.001
Cu	0.251	−0.427
Fe	0.207	0.577
Ca	−0.201	0.224
Mn	0.160	−0.032
Cd	0.645	0.254
Hg	−0.385	0.659

6.6.3 REDUNDANCY ANALYSIS

Redundancy analysis (RDA) is a technique which falls between multivariate multiple regression and canonical correlation analysis. Also known as reduced rank regression, it was featured in a recent paper in the journal *Ecotoxicology* (Van Wijngaarden *et al.* 1994). Whereas canonical correlation analysis takes two sets of variables and finds new ordination axes for both sets, RDA finds new axes for just the second set. It chooses these for maximal association with the first set of variables and not, as in canonical correlation analysis, for maximal association with linear combinations of variables from the first set.

6.6.4 CANONICAL CORRESPONDENCE ANALYSIS (CCA)

Canonical correspondence analysis (CCA) is the canonical form of CA. Just as CA can be considered the non-linear equivalent of PCA, CCA can be considered the non-linear equivalent of canonical correlation analysis. With CCA, the aim is to derive new ordination axes based on one set of variables (typically environmental variables) which give the maximum predictive agreement with ordination axes derived simultaneously from a second set of variables (typically species). The difference between CCA and canonical correlation analysis is that the latter assumes linear relationships between the two set of axes, while CCA assumes that the species have a unimodal response to the underlying environmental gradients which the ordination axes are intended to represent. As with CA, each species will have an optimum position along the gradient which can be estimated from the analysis if it occurs within the range of environmental conditions encompassed by the data set. Detrended canonical correspondence analysis (DCCA) is the canonical form of DCA, and works in the same way, by eliminating quadratic relationships between CCA axes. The CANOCO software package and user manual (Ter Braak 1988) provides a detailed introduction to CCA and other ordination techniques with worked examples.

REFERENCES

Bray RJ and Curtis JT (1957) An ordination of the upland forest communities of southern Wisconsin. *Ecological Monographs*, **27**, 325–349.

Chatfield C and Collins AJ (1980) *Introduction to Multivariate Analysis*, Chapman & Hall, London.

Clarke KR and Ainsworth M (1993) A method for linking multivariate community structure to environmental variables. *Marine Ecology Progress Series*, **92**, 205–219.

Digby PGN and Kempton RA (1987) *Multivariate Analysis of Ecological Communities*, Chapman & Hall, London.

Durell SEA Le V. dit, Goss-Custard JD and Caldow RWG (1993) Sex-related differences in diet and feeding method in the oyster-catcher *Haematopus ostralegus. Journal of Animal Ecology*, **62**, 205–215.

Gordon AD (1981) *Classification*, Chapman & Hall, London.

Gower, JC (1971) A general coefficient of similarity and some of its properties. *Biometrics*, **27**, 857–872.

Hill MO (1979) DECORANA – A FORTRAN program for detrended correspondence analysis and reciprocal averaging, *Ecology and Systematics*, Cornell University, Ithaca, New York.

Jongman FHG, Ter Braak CJF and Van Tongeren OFR (eds) (1995) *Data Analysis in Community and Landscape Ecology*, Cambridge University Press, Cambridge.

Krzanowski WJ (1988). *Principles of Multivariate Analysis, A User's Perspective*, Clarendon Press, Oxford.

Legendre L and Legendre P (1983) *Numerical Ecology*. Elsevier, Amsterdam.

Manly BFJ (1986) *Multivariate Statistical Methods. A Primer*. Chapman & Hall, London.

Manly, BFJ (1997) *Randomisation, Bootstrap and Monte Carlo Methods in Biology*, 2nd edn, Chapman & Hall, London.

Mullis, RM, Revitt DM and Shutes RBE (1996) The determination of the toxic influences to *Gammarus pulex* (amphipoda) caged in urban receiving waters. *Ecotoxicology*, **5**, 209–215.

Seber GAF (1984) *Multivariate Observations*, Wiley, New York.

Ter Braak CJF (1988) CANOCO – a FORTRAN program for canonical community ordination, Microcomputer Power, Ithaca, New York.

Van Wijngaarden RPA, Van Den Brink PJ, Oude Vashaar JH and Leeuwangh P (1994) Ordination techniques for analysing response of biological communities to toxic stress in experimental ecosystems. *Ecotoxicology*, **4**, 61–77.

Wright JF, Moss D, Armitage PD and Furse MT (1984) A preliminary classification of running-water sites in Great Britain based on macroinvertebrate species and the prediction of community type using environmental data. *Freshwater Biology*, **14**, 221–256.

Environmental Monitoring: Investigating Associations and Trends

LOVEDAY L. CONQUEST

College of Ocean & Fishery Sciences, University of Washington, USA

7.1 INTRODUCTION

This chapter addresses the topic of environmental monitoring; specifically, the issue of following some sort of process or signal over time. The word 'monitor' is related to the verb 'to warn'. Monitoring captures the idea of watching a signal over time to see if a change has occurred, which may indicate changes in pollution levels; these may correspond to changes in human activity. For example, Ratcliffe (1970) presents data concerning eggshell thickness in British birds from 1900 to 1970, along with a multi-pronged argument that decreases in eggshell thickness are likely due to the increased use of pesticides beginning in the late 1940s.

In tracking a signal over time, investigators are often interested in detecting and characterizing a trend, which, if present, may take on a variety of forms. The Ratcliffe (1970) eggshell data (updated in Newton and Haas 1984) are an example of a step change, where the general level of the time sequence shifts, presumably due to some intervention event, in this case a dramatic increase in the industrial, agricultural, and domestic use of chemical pesticides. Other trends can be more gradual, such as the build-up of a pollutant in a system over time. Many biological signals exhibit a seasonal effect, so any trend analysis must take seasonality into account. Figure 7.1 (after Gilbert 1987) displays some trends that might be encountered when engaging in environmental monitoring.

Ecotoxicology data pose special problems for the investigator. They may exhibit the following characteristics: (1) the data records may be short, thus ruling out the possibility of using sophisticated time series analyses; (2) natural background variability may be high, making it more difficult to detect trends statistically; (3) measurement techniques and sensitivities of analytical methods

Statistics in Ecotoxicology. Edited by T. Sparks. © 2000 John Wiley & Sons Ltd

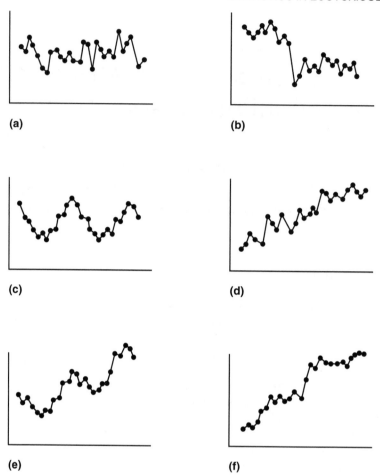

Figure 7.1 Types of trend. Vertical axis indicates the response; horizontal axis indicates the predictor variable (e.g. time). (a) random fluctuations with no trend; (b) step change; (c) seasonal pattern with no trend; (d) random fluctuations around a linear trend; (e) random fluctuations around a linear trend with seasonality; (f) correlated fluctuations around a linear trend. From R. O. Gilbert, *Statistical Methods in Environmental Pollution Monitoring*, Copyright © (1987, Van Nostrand Reinhold). Adapted by permission of John Wiley & Sons, Inc.

may have changed over time, making interpretation of detected changes more difficult (Montgomery and Reckhow 1984). Furthermore, in order to form causal explanation of trends, a knowledge of land use and other human activities is required, in addition to relevant biological process knowledge.

Most of the statistical methods discussed in this chapter are distribution-free or non-parametric, tests. (Some have been introduced in earlier chapters.) They have the advantage of not relying upon the data being normally distributed (more specifically, the error terms in the statistical model do not have to be normally distributed). Such procedures are often resistant to outliers or to statistically 'influential' data points. They also can prove useful in dealing with 'unstructured sampling', which can occur when investigators are presented with data from various sources, all of whom are tracking toxicants in birds, for example. However, non-parametric procedures are not devoid of all assumptions. Even for non-parametric regression, described later in this chapter, the usual assumption is that the error terms in the statistical model are uncorrelated.

Methods are also described that can handle correlated data, or data sets with missing data, or data that can only be measured down to a certain level (non-detects). Incorporating seasonality is also addressed. Finally, some basic concepts in dealing with time series data are introduced.

In the notation that follows, $P(1)$ and $P(2)$ refer to significance levels of one- and two-tailed tests respectively.

7.2 NON-PARAMETRIC MEASURES OF CORRELATION

Kendall's tau and Spearman's r_s are two non-parametric coefficients of association that measure the strength of a monotonic relationship between two variables, X and Y. They are non-parametric alternatives to the parametric Pearson's correlation coefficient r, which traditionally assumes a bivariate normal distribution in order to properly assess statistical significance. Both investigate the strength of the linear relationship between the ranks of X and the ranks of Y. They are resistant to unusual values in the data and can handle non-detects. Spearman's r_s gives more weight between data values ranked further apart. Helsel and Hirsch (1992) point out that the large sample approximations for r_s do not fit the distribution of the test statistic well for sample sizes < 20, in contrast to Kendall's tau; hence tau is often preferred over r_s. (Kendall's tau also generalizes to a partial correlation coefficient; see Siegel and Castellan 1988.) Spearman's r_s, however, is simply the regular Pearson correlation coefficient computed on the ranks of the X and Y; hence it is easier for some to grasp intuitively.

7.2.1 SPEARMAN'S r_s

Rank correlation refers to computing a correlation coefficient based upon the ranks of the two variables involved (call them X and Y), where each of the two variables is ranked separately. Thus if high and low values of X tend to be associated, respectively, with high and low values of Y, Spearman's r_s, the correlation coefficient based upon the ranks, should be high. Rank correlation coefficients place no specific distributional assumptions (i.e. the normal distribution) on X or Y. If X is a 'predictor' variable, like age of organism, or time, it does not have to be equally spaced. There is the assumption that X and Y are based upon underlying continuous distributions (even if they can only be measured on an ordinal basis) and that the data points are independent. Spearman's r_s is a measure of the linearity that exists between the ranks of X and Y, rather than the original values of X and Y. It therefore is a measure of the monotonicity between X and Y, and so any transformation (such as the logarithmic or square root) that preserves monotonic relationships will always yield the same value for r_s. Its ease of calculation means that rank correlation calculation is easily available on a variety of statistical software packages. If one computes the usual (parametric) Pearson correlation coefficient with the original values replaced by their ranks, after some algebraic simplification the following is obtained:

$$r_s = 1 - 6\frac{\sum\limits_{i=1}^{n} [R(X_i) - R(Y_i)]^2}{n(n^2 - 1)} \qquad (7.1)$$

where $R(X_i)$ represents the rank of the ith value of X, and $R(Y_i)$ is defined similarly.

Figures 7.2 and 7.3 compare the use of the Pearson correlation r and Spearman's r_s on data concerning toxicants in great skua eggs (Furness and Hutton 1979, Appendix 1). Figure 7.2 plots polychlorobiphenyl (PCB) concentration against hexachlorobiphenyl (HCB) concentration, both in actual values (Figure 7.2a) and in ranks (Figure 7.2b), with a least squares regression line fitted in each plot. Also in Figure 7.2a is a LOWESS (locally weighted scatterplot smoother) smoothed curve for the data (Cleveland 1979, 1981). Figure 7.2a yields a Pearson $r = 0.80$; the relationship appears monotonic but not necessarily linear. Also, there is an 'extreme' data point in the upper right-hand corner. While it does not appear to be an outlier, standard regression output from software such as MINITAB confirms that it is an influential data point; vertical shifts in its position can cause disproportionate shifts in a fitted regression line. (For more examples of this phenomenon, see Anscombe 1973 or Weisberg 1985). The plot of the ranks in Figure 7.2b is much more evenly spread out, and the earlier effect of the extreme data point has been considerably dampened. In fact, as long as that particular data point holds the highest X-value and the highest Y-value, no change in its ranks occurs. The value of Spearman's r_s for these data is 0.91, indicating the strong linear nature of the ranks. Remember, linearity in the ranks indicates monotonicity in the original values.

The plots in Figure 7.3 show PCB and 1,1-bis(4-chlorophenyl)-2,2-dichloroethane (DDD) concentration and their ranks for the great skua egg data set. Here there appears to be a general, not particularly strong, positive association between PCB and DDD. For the original values, Pearson $r = 0.31$, not significant when testing for positive association $(P(1) > 0.10)$. For their ranks, Spearman $r_s = 0.515$, yielding a significant result when testing for positive

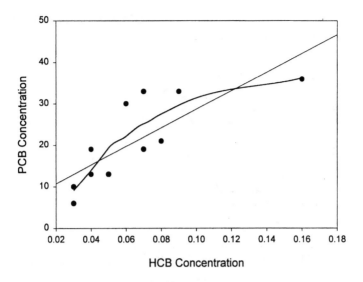

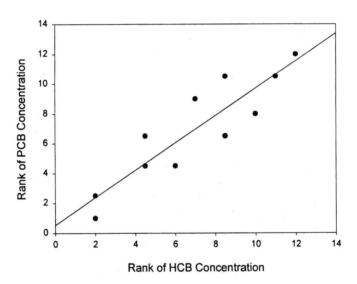

Figure 7.2 PCB concentrations versus HCB concentrations (ppm, wet weight) in great skua eggs, from Furness and Hutton (1979). (a) Actual values, including least squares regression lines and a LOWESS smoothed curve; (b) ranks, including least squares regression line on the ranks.

association ($P(1) < 0.05$). Also, the linear trend is more apparent in the plot of the ranks than in the plot of the original concentrations.

Sparks & Rothery (1996) present an example showing the association between freshwater shrimp and stickleback abundance in 54 small water bodies. Because the linear relationship is a log-log one (i.e. the linear relationship becomes evident when plotting the logarithms of the two variables against each

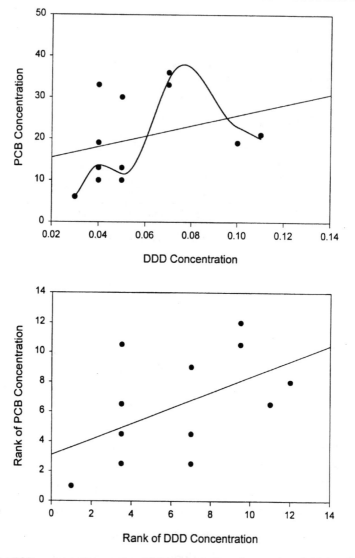

Figure 7.3 PCB concentrations versus DDD concentrations (ppm, wet weight) in great skua eggs, from Furness and Hutton (1979). (a) Actual values, including least squares regression lines and a LOWESS smoothed curve'; (b) ranks, including least squares regression line on the ranks.

other), the parametric Pearson r does not attain a statistically significant value, while the Spearman r_s does. Furthermore, the Spearman r_s maintains the same value, whether one is considering the original variables, or taking the logarithm of one or both of them. Using ranks tends to modify (by dampening) the effect of 'influential' data points, data points that live on the edges of a scatterplot, far from the centre of the data, that can disproportionately pull a least squares fitted line up or down and thus have undue influence in the assessment of linear relationships. One therefore does not have to worry as much about outliers when using rank correlation.

Rank correlation has other advantages, one being that it will handle data consisting of 'non-detects' (where all one knows is that the value is less than some number, or perhaps the value is recorded as a 'trace'). Similarly, rank correlation will handle right-censored data, where a value is recorded as being greater than some number. Since toxicology data sometimes exhibit non-detects, this is a useful property. At the same time, it must be noted that some information is always lost when replacing actual observations by their ranks. Sparks and Rothery (1996) also present some cases with non-normal data where the Pearson r is significant (or nearly so) but the Spearman r_s is quite non-significant. In these cases, a significance p-value based upon random reorderings of the original data (and thus distribution-free) gives more powerful results than simply using the rank correlation.

7.2.2   KENDALL'S TAU

Kendall's tau is another measure of the strength of the monotonic relationship between two variables. Like the Spearman rank correlation coefficient, it is resistant to unusual values and can handle non-detects. Kendall's tau is especially good for assessing monotonic relationships that are non-linear; it does assume independent data points. Helsel and Hirsch (1992) note that it works well for variables having skewed distributions around the general relationship. Tau has advantages over r_s in that it generalizes easily to accommodate multiple stations and seasonality.

To compute Kendall's tau, first order the data points by values of increasing X (repeat values of X are allowed). For example, if X is time, then the data points would be ordered respectively according to increasing values of time. (An advantage here is that the time intervals would not necessarily have to be equally spaced.) In the presence of positive correlation, the Y-values will tend to increase (more often than decrease) as X increases. The opposite will happen when correlation is negative. In order to compute tau, one counts the number of 'concordant pairs' (pairs where y increases with increasing X) and the number of 'discordant pairs' (pairs where Y decreases with increasing X). This process is illustrated in Table 7.1 on more great skua contamination data from Furness and Hutton (1979). This particular set of data relates the age of great skuas to cadmium concentrations in kidney tissues. Note that concordant pairs are marked with a plus sign and discordant pairs are marked with a minus sign. A pair where the two Y-values match exactly is accorded a zero. Also, if two or more data points share the same X-value, then no comparison is done between their Y-values. Finally, the difference between the number of concordant pairs and discordant pairs is compared to the $n(n-1)/2$ possible comparisons that

Table 7.1 Computation of Kendall's tau for age and cadmium levels in great skua kidney tissues

Age (x)	3	4	4	5	5	7	7	8	8	9	10	12	No. of + signs	No. of − signs
Cadmium level (y)	64.6	13.5	50.0	14.5	52.6	21.6	129.2	29.7	58.6	114.7	91.2	336.0	+ signs	− signs
		−	−	−	−	−	+	−	−	+	+	+	4	7
			NC	+	+	+	+	+	+	+	+	+	9	0
				−	+	−	+	−	+	+	+	+	6	3
					NC	+	+	+	+	+	+	+	7	0
						−	+	−	+	+	+	+	5	2
							NC	+	+	+	+	+	5	0
								−	−	−	−	+	1	4
									NC	+	+	+	3	0
										+	+	+	3	0
											−	+	1	1
												+	1	0
													45	17

$$\text{tau} = \frac{45 - 17}{\dfrac{12 \cdot 11}{2}} = .42$$

Note that the x-values are ordered. Each y measurement is compared to the measurement appearing to the right of it. A '+' denotes that the sign of the difference of the y-values is the same as the sign of the difference of the x-values; a '−' denotes the opposite relationship. When multiple x-values occur, the y-values are not compared (NC). See text for Mann–Kendall test of significance.

could be made among the n data pairs. Kendall's tau will always assume a value between -1 (complete discordance; Y decreases as X increases) and $+1$ (complete concordance; Y increases as X increases). Then, Kendall's tau correlation coefficient is

$$\frac{(\text{no. of concordant pairs} - \text{no. of discordant pairs})}{\dfrac{n(n-1)}{2}} \qquad (7.2)$$

The statistical significance of tau is assessed by performing a test on the numerator, $S = $ no. concordant pairs $-$ no. discordant pairs. This test, the Mann–Kendall test, is described in the next section.

7.3 THE MANN–KENDALL TEST

The Mann–Kendall non-parametric test for trend (or more generally, for association between two variables) is based upon using the Kendall's tau correlation coefficient that was introduced above (Mann 1945; Kendall 1975). The test is distribution-free and allows both non-detects and missing values; it does assume that the error terms are independent (Hollander and Wolfe 1973). It uses only the relative magnitudes (ranks) of the data rather than the actual values. If one of the two variables is time, then the Mann–Kendall test can be viewed as a non-parametric test for zero slope of the linear regression of time-ordered data against time. The null hypothesis is that there is no trend of the Y-variable when plotted against the X-variable, versus the alternative that a trend exists. The alternative hypothesis may be two-sided (i.e. if a trend is present, it could be either increasing or decreasing) or one-sided (the investigator has prior reason to favour a trend in a particular direction). The actual value of the test statistic is based upon the numerator S of the Kendall correlation coefficient; that is,

$$S = \text{no. of concordant pairs} - \text{no. of discordant pairs} \qquad (7.3)$$

Tabled values for S, with exact P-values for up to $n = 40$ are given, for example, in Table A.21 in Hollander and Wolfe (1973), or Kaarsemaker and van Wijngaarden (1953). For $n > 40$, a normal approximation may be used; this approximation is reputed to be quite good for n as low as 10 unless there are many tied X- or Y-values (Kendall 1975; Gilbert 1987). The variance of S (with a correction for tied values) is computed as follows:

$$\text{Var}(S) = \frac{1}{18}\left[n(n-1)(2n+5) - \sum_{p=1}^{g} t_p(t_p - 1)(2t_p + 5) - \sum_{q=1}^{h} t_q(t_q - 1)(2t_q + 5) \right]$$
$$(7.4)$$

where g (and h) is the number of tied Y-groups (tied X-groups) and t_p (and t_q) is the number of tied values in the pth (qth) group. Then S and its estimated variance, Var(S) are used as follows to compute an approximately normal Z-statistic:

$$Z = \frac{(S-1)}{\sqrt{\text{Var}(S)}} \quad \text{if } S > 0$$

$$= \frac{(S+1)}{\sqrt{\text{Var}(S)}} \quad \text{if } S < 0 \tag{7.5}$$

$$= 0 \quad \text{if } S = 0.$$

For tied observations in the X-variable (such as multiple observations per time period), Gilbert (1987, p. 213) gives a longer, more accurate formula for the variance of S. Fortunately for the user, for most non-parametric tests, corrections for ties tend to result in very small changes in the value of the test statistic. When there are multiple observations per time period, one should check to see whether these observations are likely to be correlated. (For example, with the skua cadmium data in Table 7.1, there are four groups of birds with the same age, although there is no reason to believe for these data that the birds are not independent.) If the observations are correlated (e.g. measuring a toxicant on birds from the same nest, or measuring eggshell thickness on eggs from the same clutch), it may be preferable to compute a summary statistic such as a median or mean for each time period, and then apply the Mann–Kendall test using the independent summary statistics.

Let us do a test of significance on the value of Kendall's tau of 0.42 which was computed for the association of age and cadmium levels in great skua kidney tissues. There are no tied Y-values, but there are multiple observations per time period. We will do the approximate Z-test both with and without the variance correction for ties and compare them to exact results from Table A.21 in Hollander and Wolfe (1973).

Numerator of the Z-statistic = 28 − 1 = 27. No correction for ties:

$$\text{Var}(S) = \frac{12(11)(2 \cdot 12 + 5)}{18} = 212.67$$

$$\sqrt{212.67} = 14.58 \tag{7.6}$$

$$Z = 1.85$$

Correcting for the four groups of tied X − values (four groups with two ties each):

$$\text{Var}(S) = \frac{12(11)(2 \cdot 12 + 5) - (18 + 18 + 18 + 18)}{18} = 208.67$$

$$\sqrt{208.67} = 14.44 \tag{7.7}$$

$$Z = 1.87$$

In testing the null hypothesis that tau = 0 against a one-sided alternative that the cadmium levels tend to be larger in older birds, $P(1) = 0.032$ without correcting for ties and 0.031 when correcting for ties. Hollander and Wolfe's Table A.21 (1973) give the exact P-value as 0.031. So even without correcting for ties, one comes quite close to the exact P-value.

7.4 THE MANN–KENDALL TEST WITH NON-DETECTS

Hughes and Millard (1988) have extended the Mann–Kendall test to cover the situation where there are multiple non-detects. They propose a method that is more powerful than simply ranking all values less than the highest censoring level as tied. Consider the following data set where '$<$' denotes a non-detect value:

Time i	1	2	3	4	5
Observation X_i	<3	13	10	<11	15
Possible rank vectors					
	1	4	2	3	5
	1	4	3	2	5
	2	4	3	1	5
Expected rank vector	1.3	4	2.7	2	5

In order to tell whether a pair of values is concordant or discordant, we must be able to rank the observations. However, it is not possible to tell whether the true (but unknown) value of X_4 is greater than or less than X_3, or even X_1. Because X_2 and X_5 always have the top two ranks, and because $X_1 < X_3$, there are three possible rank vectors; these are listed along with the data. To resolve the ambiguity in ranks, the average rank is computed for each observation, yielding the 'expected rank vector'. Then Kendall's tau and the Mann–Kendall test are computed based upon this expected rank vector. The variance for the new test statistic, S_{erv} (for 'expected rank vector'), is the standard formula to correct the variance for ties in the Y-values; the correction for ties would be used whenever ties occur in the expected rank vector. This does assume that the mechanism of censoring is independent of time. If this is not the case (e.g. if the method of measurement improves over time, thus causing the change in detection values), then the method of figuring out all possible rank vectors becomes much more complex (see Hughes and Millard 1988).

For the simple data set presented here, using the expected rank vector, there are six concordant pairs and four discordant pairs, for a difference of 2. Thus, Kendall's tau is equal to $(6 − 4)/10 = 0.20$.

To do the Mann–Kendall test, the computed variance is

$$\frac{5(5 − 1)(2 \cdot 5 + 5)}{18} = 16.67 \tag{7.8}$$

Table 7.2 Computation of Theil slope estimator (ranks in parentheses)

j:	1	2	3	4	5	6	7	8	9	10	11	12
x:	3	4	4	5	5	7	7	8	8	9	10	12
y:	64.6	13.5	50.0	52.6	14.5	129.2	21.6	29.7	58.6	114.7	91.2	336.0
$(y_j - 64.6)/(x_j - 3.),\ j > 1$		−51.1 (3)	−14.6 (8)	−6.0 (15)	−25.5 (5)	16.15 (37)	−10.75 (10)	−6.98 (14)	−1.2 (17)	8.35 (30)	3.8 (24)	30.15 (43)
$(y_j - 13.5)/(x_j - 4.),\ j > 2$			NC	39.1 (49)	1.0 (18)	38.57 (48)	2.7 (22)	4.05 (25)	11.27 (31)	20.24 (39)	12.95 (33)	40.31 (50)
$(y_j - 50.0)/(x_j - 4.),\ j > 3$				2.6 (21)	−35.5 (4)	26.4 (42)	−9.47 (11)	−5.07 (16)	2.15 (20)	12.94 (32)	6.87 (27)	35.75 (45)
$(y_j - 52.6)/(x_j - 5.),\ j > 4$					NC	38.3 (47)	−15.5 (7)	−7.63 (12)	2.05 (19)	15.52 (36)	7.72 (28)	40.49 (51)
$(y_j - 14.5)/(x_j - 5.),\ j > 5$						57.35 (56)	3.55 (23)	5.07 (26)	14.7 (34)	25.05 (41)	15.34 (35)	45.93 (53)
$(y_j - 129.2)/(x_j - 7.),\ j > 6$							NC	−99.5 (1)	−70.6 (2)	−7.25 (13)	−12.67 (9)	41.36 (52)
$(y_j - 21.6)/(x_j - 7.),\ j > 7$								8.1 (29)	37.0 (46)	46.55 (54)	23.20 (40)	62.88 (57)
$(y_j - 29.7)/(x_j - 8.),\ j > 8$									NC	85.0 (61)	30.75 (44)	76.57 (60)
$(y_j - 58.6)/(x_j - 8.),\ j > 8$										56.10 (55)	16.30 (38)	69.35 (58)
$(y_j - 114.7)/(x_j - 9.),\ j > 10$											−23.50 (6)	73.77 (59)
$(y_j - 91.2)/(x_j - 10.),\ j > 11$												122.40 (62)

Data are cadmium concentrations (ppm) in great skua kidneys, versus age (years), from Furness and Hutton (1979). 'NC' means no comparison possible due to repeat age values. There are 62 slope values, so median slope = average of slopes with ranks 31 and 32 = (11.27 + 12.94)/2 = 12.10.

This yields an approximate Z-statistic of

$$\frac{(2 - 1)}{\sqrt{16.67}} = 0.24 \tag{7.9}$$

with a one-sided P-value of 0.40.

Consulting Table A.21 of Hollander and Wolfe (1973), we find that the exact P-value is 0.408. This method using the expected rank vector may also be extended to the seasonal Mann–Kendall test (which appears later in this chapter) by computing the expected ranks within seasons.

7.5 ESTIMATING TREND; MEDIAN SLOPE REGRESSION

When an investigator concludes that trend is present, then it becomes relevant to quantify the trend. The simplest parametric approach involves computing the value of the estimated linear slope coefficient using simple linear regression. However, a slope computed in this manner can deviate greatly from the true slope due to the presence of outliers, high variability or if the relationship is monotonic but not linear. The following non-parametric slope estimator was proposed by Theil (1950) and extended to incorporate repeat X-values by Sen (1968). It is not affected by outliers and can be computed when data are non-detects or missing. Conceptually, one computes the slope for each pair of data points; the Theil estimator is defined to be the median of all these slopes. More formally, then, the Theil estimator is defined as follows:

$$b_1 = \text{median} \left\{ \frac{(Y_j - Y_i)}{(X_j - X_i)} \right\} \quad \text{for all } i < j, \, i = 1, 2, 3, \ldots, n - 1 \tag{7.10}$$

$$j = 2, 3, \ldots, n$$

If N, the number of computable slopes, is odd, then the median will simply be the middle value of the slopes. If N is even, then the median is defined to be the average of the two middle values. If all the X-values are distinct and there are no missing data or non-detects, then there will be $n(n - 1)/2$ computed slopes, and b_1 is defined as the median slope. If there are repeated X-values, or if some of the data points have missing values or are non-detects, then the number of computed slopes will be fewer than $n(n - 1)/2$. Once the median slope b_1 has been found, then the intercept is defined as

$$b_0 = \text{median} \{Y\text{'s}\} - b_1 \text{ median} \{X\text{'s}\} \tag{7.11}$$

A two-sided confidence interval for the Theil slope estimator, based on a normal approximation is also available (Hollander and Wolfe 1973). First, choose the appropriate Z-value that corresponds to the desired level of confidence (e.g. 1.96 for a 95% confidence interval, 1.645 for 90%, 1.28 for 80%). Then, compute

$$C(\alpha) = Z \cdot \sqrt{\text{Var}(S)} \tag{7.12}$$

taking the closest integer. Set

$$M_1 = \frac{N - C(\alpha)}{2} \qquad M_2 = \frac{N + C(\alpha)}{2} \qquad (7.13)$$

Then the lower and upper limits of the confidence interval are the M_1th largest and the $(M_2 + 1)$th largest of the N ordered slopes.

This procedure is illustrated on the skua data in Table 7.2. All 62 possible 'two-point slopes' are computed, and the median slope is found to be equal to 12.10. Then the intercept is

$$
\begin{aligned}
b_0 &= \text{median}\,\{Y\text{'s}\} - 12.10 \cdot \text{median}\,\{X\text{'s}\} \\
&\quad 55.6 - 12.10 \cdot 7.0 = -29.1
\end{aligned} \qquad (7.14)
$$

Running an ordinary least squares regression to the same data set yields the fitted equation,

$$\text{cadmium concentration} = -73.52 + 22.66 \cdot \text{age} \qquad (7.15)$$

Hence the least squares regression results in a steeper line with a lower intercept. Figure 7.4 reveals how much the ordinary least squares fit is influenced by a single data point in the upper right-hand corner; this data point is tagged as being 'unusual' since its standardized residual exceeds 2.0. Median slope regression is much less influenced by influential data points like these.

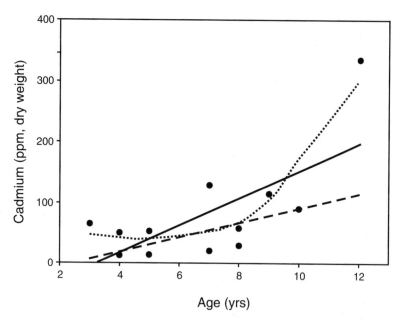

Figure 7.4 Cadmium concentrations (ppm, dry weight) versus age (years) in great skuas, from Furness and Hutton (1979). Solid line is the least squares regression lines. Dashed line is the median slope regression line. Dotted curve is the LOWESS smoothed curve.

We earlier computed Var(S) for these data as 208.67, correcting for ties in age. For a 90% confidence interval, we use $Z = 1.645$.

$$C(0.05) = 1.645 \cdot \sqrt{208.67} = 23.8 \tag{7.16}$$

with the closest integer being 24.

$$M_1 = \frac{(62 - 24)}{2} = 19 \tag{7.17}$$

$$M_2 = \frac{(62 + 24)}{2} = 43 \tag{7.18}$$

So, the lower and upper 90% confidence interval limits are the 19th largest (slope value = 2.05) and the 44th largest (slope value = 30.75) of the 62 ordered slopes. Ordinary least squares regression yields a 90% confidence interval (based on 1.812 = t-value for 10 d.f. and a standard error of the slope of 7.34) of (9.36, 35.96), somewhat shorter in length but again influenced by the extreme data point.

7.6 THE SEASONAL KENDALL TEST AND SEASONAL KENDALL SLOPE ESTIMATOR

When seasonal cycles are present in the data, then it is necessary to use a test that either accounts for them or is not affected by them. The Mann–Kendall test presented earlier has been generalized to the seasonal Kendall test (Hirsch *et al.* 1982; Smith *et al.* 1982; van Belle and Hughes 1984). Conceptually, the Mann–Kendall test statistic S and its variance, Var(S), are computed separately for each of m seasons, and the results are combined at the end. Thus one avoids making any comparisons across seasonal boundaries.

To carry out the actual test, first compute the Mann–Kendall statistic S_i for each season, along with its variance, Var(S_i). Then obtain the overall statistic S' by summing up the separate values of S_i across all the seasons. Similarly, sum the variances, Var(S_i), across all the seasons. We then rely upon an approximate Z-statistic as follows:

$$
Z = \frac{(S' - 1)}{\sqrt{\mathrm{Var}(S')}} \quad \text{if } S' > 0
$$
$$
0 \qquad\qquad \text{if } S' = 0 \tag{7.19}
$$
$$
= \frac{(S' + 1)}{\sqrt{\mathrm{Var}(S')}} \quad \text{if } S' < 0
$$

The seasonal Kendall slope estimator is also a generalization of the Theil–Sen slope estimator. First, N_i individual slope estimates for the ith season are computed as before, remembering to look at the data for each season

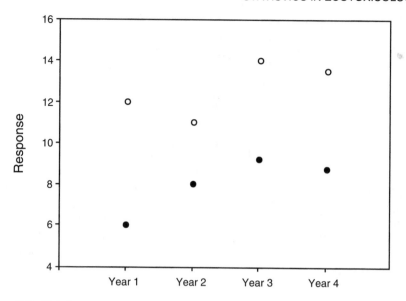

Figure 7.5   Plot of a response variable against time, for each of two seasons, denoted by open circles and solid circles.

separately. Then rank all of the individual slope estimators (now combined into a single group, say there are N' of them) and find their median. This median is the seasonal Kendall slope estimator. A 95% confidence interval is found the same way as in the previous section, now using N' slope estimators:

1. Choose the Z-value that corresponds to the desired level of confidence.
2. Compute $C(\alpha) = Z \cdot \sqrt{\text{Var}(S')}$, taking the closest integer.
3. Set

$$M_1 = \frac{(N' - C(\alpha))}{2} \quad \text{and} \quad M_2 = \frac{(N' + C(\alpha))}{2} \qquad (7.20)$$

4. The lower and upper limits of the confidence interval are the M_1th largest and the $(M_2 + 1)$th largest of the N' ordered slope estimates.

An example with a small, artificial data set is illustrated in Table 7.3 and Figure 7.5. Note that the plot shows that the data from season 2 are generally higher than season 1 (e.g. season 1 might be the dry season and season 2 the wet one); both seasons appear to show a slight upward linear trend. Although the seasonal Kendall slope turned out to be not significantly different from zero, for the purpose of illustration let us go ahead and do the 95% confidence interval for the slope. Using $Z = 1.96$ for 95% confidence,

$$C(0.05) = 1.96 \cdot \sqrt{\text{Var}(S')} = 1.96 \cdot \sqrt{17.33} = 8.16 \qquad (7.21)$$

Table 7.3 Computation of seasonal Kendall slope estimator, seasonal Kendall test, and homogeneity of seasons test

Season 1

Year	1	2	3	4	No. of + signs	No. of - signs
Data	6.0	8.0	9.2	8.7		
Slopes		2.0	1.6	0.9	3	0
			1.2	0.35	2	0
				-0.5	0	1
					$S_1 =$ 5	- 1
						= 4

Season 2

Year	1	2	3	4	No. of + signs	No. of - signs
Data	12.0	11.0	14.0	13.5		
Slopes		-1.0	1.0	0.5	2	1
			3.0	1.25	2	0
				-0.5	0	1
					$S_2 =$ 4	- 2
						= 2

Each slope is $(y_j - y_i)/(year_j - year_i)$ for $i > j$

Ordered slopes: $-1.0, -0.5, -0.5, 0.35, 0.50, 0.90, 1.0, 1.2, 1.25, 1.6, 2.0, 3.0$.

Seasonal slope estimate = median slope = average of slopes with the two middle ranks (since n is even) = $(.9 + 1.0)/2 = 0.95$.

Seasonal Kendall test on $S' = S_1 + S_2 = 6.0$ $n_1 = n_2 = 4$

Var $(S_i) = (1/18)[n_i(n_i - 1)(2n_i + 5)] = (1/18)[4(3)(13)] = 8.67$ $i = 1,2$

Var $(S') = $ Var $(S_1) + $ Var $(S_2) = 17.33$

Since $S' > 0$, the test statistic is $Z = (S' - 1)/\sqrt{\text{Var}(S')} = (6 - 1)/\sqrt{17.33} = 1.20$, not significant at the one-sided 10% level. In doing a test for positive slope, the one-sided P-value is $P(1) = 0.12$.

Homogeneity of seasons test ($m = 2$ seasons)

$Z_1 = S_1/\sqrt{\text{Var}(S_1)} = 4/\sqrt{8.67} = 1.36$

$Z_2 = S_2/\sqrt{\text{Var}(S_2)} = 2/\sqrt{8.67} = 0.68$

$\bar{Z} = (Z_1 + Z_2)/2 = 1.02$

Homogeneity test statistic $= (Z_1^2 + Z_2^2) - m\bar{Z}^2 = 1.36^2 + 0.68^2 - 2(1.02^2) = 0.23$

Compare to a chi-square distribution with $m - 1 = 1$ d.f.; P-value is $0.50 < P < 0.75$, a non-significant result (i.e. data are consistent with null hypothesis of homogeneity of seasons).

so the closest integer is 8.

$$M_1 = \frac{12 - 8}{2} = 2 \qquad M_2 = \frac{12 + 8}{2} = 10 \qquad (7.22)$$

The lower and upper limits of the confidence interval are the second largest and the eleventh largest of the 12 order slope estimates, so the 95% confidence interval is $(-0.5, 2.0)$, which overlaps zero and corresponds to the non-significant result that we just concluded.

For comparison, a least squares linear regression of response against year, with a dummy variable for season, yields a slope of 0.84 with a standard error of 0.30, significant at the 0.05 level ($P(2) < 0.05$). In this case the lower power of the non-parametric approach (assuming that the data are indeed normally distributed) has made it harder to attain statistical significance.

This method assumes independence among the seasonal test statistics. If serial correlation is suspected, then the significance test can be modified to accommodate this (Hirsch and Slack 1984). However, the authors recommend having at least 10 years of monthly data for the test to be valid.

7.7 TEST FOR HOMOGENEITY OF DIFFERENT LABS, STATIONS OR SEASONS

It is quite possible for environmental monitoring data to arise from different locations, laboratories, or stations. This raises the question of homogeneity of trends across the stations. Van Belle and Hughes (1984) present a chi-square test to test for homogeneity of trend across stations. Then, if the trends are homogeneous, the general trend can be estimated. This test can also be extended to incorporate seasonality.

For each of m stations, compute the Z-statistic,

$$Z_i = \frac{S_i}{\sqrt{\text{Var}(S_i)}} \qquad (7.23)$$

where S_i is the Mann–Kendall $S-$ statistic for the ith station. Sum the squares of the Z_i for all m stations to compute the 'total' chi-square statistic:

$$\chi^2 \text{ (total)} = \sum_{i=1}^{m} Z_i^2 \qquad (7.24)$$

Then, compute the 'trend' and 'homogeneous' chi-square components:

$$\chi^2 \text{ (trend)} = m \cdot \bar{Z}^2 \qquad (7.25)$$

where $\bar{Z}$ is the average of the Z_i's.

$$\chi^2(\text{homogeneity}) = \chi^2(\text{total}) - \chi^2(\text{trend}) \qquad (7.26)$$

The null hypothesis is that the stations are homogeneous with respect to trend. In such circumstances, the values of Kendall's tau should be similar, and the values of the Z_i's should also be similar. In that case, the expression $\sum_{i=1}^{m}(Z_i - \bar{Z})^2$ (which, after algebraic simplification, is equal to the equation for χ^2 homogeneity) as expressed above) should behave like a chi-square distribution with $m - 1$ d.f. The last part of Table 7.3 illustrates the homogeneity test for the two seasons, with a non-significant result, thus allowing one to conclude that the data are consistent with the null hypothesis of homogeneity of seasons.

7.8 MODIFICATIONS FOR MISSING TIME POINTS

Alvo and Cabilio (1993, 1994, 1995) have devised modifications for Spearman's r_s and Kendall's tau when data are missing for one or more time periods. This is useful when a sampling design has called for data to be taken at equally spaced time points, but the final data set is missing some observations. (However, the test statistics themselves, being based upon ranks, do not depend upon equal spacing of time points for the validity of the test statistics.) The modified test statistics may be considered as analogies to the situation where ties occur in a complete ranking, and the resulting test statistic is conditional on the pattern of the ties. For Spearman's rank correlation, the modified test statistic is arrived at by replacing missing ranks by the average rank for the recorded observations.

For testing the significance of Spearman's r_s, the formal test statistic is defined in the following manner:

$$A_s = \left(\frac{n+1}{k+1}\right) \cdot \sum_{i=1}^{k}\left(t_i - \frac{n+1}{2}\right)\left(R(t_i) - \frac{k+1}{2}\right) \tag{7.27}$$

where n is the total number of time points, k the number of time points with recorded observations and $R(t_i)$ the rank of the observation at time point t_i. In its most easily calculated algebraic form, the modified Spearman statistic may be expressed as

$$A_s = \left(\frac{(n+1)}{(k+1)}\right) \cdot \left(V - \frac{n(n+1)(k+1)}{4}\right) \tag{7.28}$$

where V is the product between the vector $(1, 2, \ldots, n)$ and the vector of ranks formed by putting $R(t_i)$ in position t_i for the recorded observations and $(k+1)/2$ in the remaining positions with the missing observations, thus replacing missing observations by the average rank of the recorded observations.

For example, a data vector of five time points with one missing observation at time t_2 might be, in terms of ranks: 1, [missing], 2, 4, 3. To obtain V, the missing slot is replaced by 2.5, which is the average value of the observed ranks. Then

$$V = 1 \cdot 1 + 2 \cdot 2.5 + 3 \cdot 2 + 4 \cdot 4 + 5 \cdot 3 = 43 \tag{7.29}$$

and

$$A_s = \left(\frac{(5 + 1)}{(4 + 1)}\right) \cdot \left(43 - \frac{5(5 + 1)(4 + 1)}{4}\right) = 6.6 \qquad (7.30)$$

Under the null hypothesis of no trend, A_s is symmetric with mean 0. The formula for the variance depends upon whether one is concerned with where the missing values occur (conditioning upon the specific times), or whether one wishes to consider the missing slots as being randomly distributed. Under the former, the variance is as follows:

$$\text{Var}_1(A_s) = \frac{k(n + 1)^2}{(k + 1)} \cdot \sum_{i=1}^{k} \frac{(t_i - \bar{t})^2}{12} \qquad (7.31)$$

Under the latter, the variance is

$$\text{Var}_2(A_s) = \frac{k(k - 1)}{(k + 1)} \cdot \frac{n(n + 1)^3}{144} \qquad (7.32)$$

Suppose a data set consisting of 20 time points is missing the fifth observation, such that the ranks appear as: 5, 8, 13, 3, [missing], 1, 9, 7, 2, 11, 4, 15, 12, 6, 19, 16, 17, 18, 10, 14. The missing slot is replaced by the value 10, the average value of the other 19 ranks.

$$V = 1 \cdot 5 + 2 \cdot 8 + 3 \cdot 13 + 4 \cdot 3 + 5 \cdot 10 + \ldots + 20 \cdot 14 = 2463 \qquad (7.33)$$

$$A_s = \left(\frac{21}{20}\right) \cdot \left[2463 - \frac{20(21)(20)}{4}\right] = 381.15 \qquad (7.34)$$

If we condition on the observed pattern of missing observation 5, then

$$\text{Var}_1(A_s) = \frac{19}{20}(21^2)(52.76) = 22105.14 \qquad (7.35)$$

(Note that for the last summation in computing $\text{Var}_1(A_s)$, the t_i-values used were those for the observed values, namely 1, 2, 3, 4, 6, 7, 8, 9, . . ., 19, 20.) Computing an approximately normal Z-statistic, then

$$Z = \frac{A_s}{\sqrt{\text{Var}_1(A_s)}} = \frac{381.15}{\sqrt{22105.14}} = 2.5635 \qquad (7.36)$$

statistically significant with a two-sided P-value of about 0.01.

If we condition on the fact that there is one missing observation (which could have occurred in any one of the 20 time periods), then

$$\text{Var}_2(A_s) = \frac{19(18)}{20} \cdot \frac{20(21^3)}{144} = 21994.875 \qquad (7.37)$$

Then the Z-statistic is

$$Z = \frac{A_s}{\sqrt{\text{Var}_2(A_s)}} = \frac{381.15}{\sqrt{21994.875}} = 2.57 \qquad (7.38)$$

statistically significant with a two-sided P-value of 0.01.

For testing the significance of the Kendall tau correlation coefficient, the formal test statistic is defined as follows: with the time vector defined as $(1, 2, \ldots, n)$, we need to consider 'score functions' $a(i, j)$ for every pair of time points, $1 \leq i < j \leq n$, defined as

$a(i, j) = 1 \cdot \text{sign}(R(i) - R(j))$ if both time points are observed

$\dfrac{2R(j)}{(k+1)} - 1$ if the earlier time point is missing but not the later

$1 - \dfrac{2R(i)}{(k+1)}$ if the later time point is missing but not the earlier

0 if both time points are missing   (7.39)

Then the test statistic

$$A_k = \sum_{i<j} a(i, j) \qquad (7.40)$$

What this version of the Kendall tau does is to again subtract the number of discordant pairs from the number of concordant pairs as before, with some added corrections for any pair that includes a missing value.

For example, for the above vector of 20 points through time with the missing observation at time t_5, there are 117 concordant pairs and 56 discordant pairs, counted in the same manner as illustrated in Table 7.1. The only pairs of time points that involve computation of $a(i,j)$ with the missing observations are time points t_1, t_2, t_3, t_4, and t_5.

For t_1, $a(1, 5) = 0.5$
For t_2, $a(2, 5) = 0.2$
For t_3, $a(3, 5) = 0.3$ (7.41)
For t_4, $a(4, 5) = 0.7$
For t_5, $a(5, j) = \dfrac{2R(j)}{20} - 1$

Summing these terms up over the 15 terms for $j = 6, 7, \ldots, 20$, yields the following:

$$\left(\frac{2}{20}\right) \cdot \sum_{j=6}^{20} R(j) - 15 = (0.1)(161) - 15 = 1.1 \qquad (7.42)$$

Hence

$$A_k = (117 - 56) + (0.5 + 0.2 - 0.3 + 0.7 + 1.1) = 63.2 \qquad (7.43)$$

We compute the variance two ways: first, assuming that the missing observation is random (that is, conditioning on the fact that there is one missing observation, which could have occurred anywhere). In that case,

$$\text{Var}_2(A_k) = \frac{k(k-1)}{9(k+1)} \cdot \left[(n + 1.5)^2 + \frac{k}{2} + 0.25\right] \qquad (7.44)$$

so for these data

$$\text{Var}_2(A_k) = \frac{19(18)}{9(20)} \cdot \left[21.5^2 + \frac{19}{2} + 0.25\right] = 896.8 \qquad (7.45)$$

Then,

$$Z = \frac{A_k}{\sqrt{\text{Var}(A_k)}} = \frac{63.2}{\sqrt{896.8}} = 2.11 \qquad (7.46)$$

statistically significant with a two-sided P-value of 0.035. Exact critical values for $n < 17$ are available in Table 4 of Alvo and Cabilio (1993).

If we condition on the fact that there is one missing observation at time t_5, then

$$\text{Var}(A_k) = \frac{16}{(n + 1)^2} \cdot \text{Var}_1(A_s) + \frac{5k(k - 1)}{18} + \frac{8}{3(k + 1)} \cdot \sum_{i=1}^{k} (t_i - i)\left(i - \frac{(k + 1)}{2}\right) \qquad (7.47)$$

which for these data = 983.21. Then

$$Z = \frac{63.2}{\sqrt{983.21}} = 2.02 \qquad (7.48)$$

statistically significant with a two-sided P-value of 0.04. Exact critical values for $n < 14$ are available in Table of Alvo and Cabilio (1993).

Regarding which variance formula to choose (fixed or random patterns of missing observations), this author surmises that most time sequences with missing observations would arise from unanticipated events (e.g. malfunction of recording instruments or unavailability of samplers). There do, however, exist documented cases where, due to high sampling and chemical analysis costs in monitoring programs, certain time periods are deliberately skipped in an otherwise equally spaced time series (e.g. Shelly 1994).

7.9 SERIAL CORRELATION

All of the tests that have been discussed thus far, while being distribution free, still assume independent data points. Particularly with data points that are taken through time, this independence cannot always be guaranteed. Sometimes each data point is influenced by the value of the previous data point, such as measuring a growth variable on the same organism through time. This phenomenon is known as autocorrelation, or serial correlation, and refers to the self-similarity of variates adjacent in space or time (Sokal and Rohlf 1995). The presence of autocorrelation means that, even after having accounted for major trends and seasonality, there is still some relation between a data point and its adjacent neighbours.

One way to check for serial correlation after having run a regression model on time-sequenced data is to compute the Durbin–Watson statistic (Durbin and Watson 1951), which is related to the serial correlation coefficient:

$$d = \frac{\sum\limits_{i=2}^{n}(e_i - e_{i-1})^2}{\sum\limits_{i=1}^{n}e_i^2} \qquad (7.49)$$

If serial correlation is present, values of adjacent residuals will tend to be similar. Hence, the value of d will tend to be too low. Tables of d are found in Rohlf and Sokal (1995) and in many time-series texts; the Durbin–Watson statistic is also available in software such as MINITAB. When $n > 25$ we can use the normal approximation

$$Z = \frac{\left| 1 - \dfrac{d}{2} \right|}{\sqrt{\dfrac{n-2}{n^2-1}}} \qquad (7.50)$$

(Sokal and Rohlf 1995).

The presence of serial correlation can also be assessed non-parametrically using Kendall's tau or Spearman's r_s. This must be done after any major trends or seasonal cycles have been removed via an appropriate regression model. After the regression model has been fitted, the resulting residuals from the model are ordered by time, and the 'lagged' residual pairs (e_i, e_{i-1}) are formed. Then Kendall's tau or Spearman's r_s may be computed for the residual pairs. If the correlation is significantly different from zero, then serial correlation is present.

Then one must face the issue of what to do in the presence of serial correlation. One possibility is to sample from the original data set in the hope that the resulting, smaller, data set will have little or no serial correlation. Since serial correlation means that there is considerable redundancy in the information, the claim is that not much information is lost by doing this (Helsel and Hirsch 1992). However, if the data set is small to begin with, then serial correlation must be accounted for in the actual test itself.

7.10　PAGE'S TEST FOR TREND WITHIN STATIONS OR SUBJECTS

Sometimes we have data taken through time on the same subject; the subject may be an organism or a group of related organisms (such as a family of animals), or an instrument placed at a particular location. Page's test (Page 1963, see also Siegel and Castellan, 1988) is set up to test for a trend through time, keeping in mind that for a given subject, measurements through time are expected to be correlated. Because it is a non-parametric test based on ranks, it does not involve trying to model or estimate the correlation structure; thus it is easy to apply. It does handle non-detects, although it does not accommodate missing values. Page's trend test is most readily described through an example.

Suppose there are four stations where a reading has been taken each year for six years. The null hypothesis would be that there is no trend over the six years; the alternative is that there has been an increasing trend. The data are as follows:

	YEAR					
	1	2	3	4	5	6
STATION						
A	8.6	8.4	6.9	10.2	9.3	10.0
B	7.4	9.6	10.2	11.3	11.5	11.5
C	8.6	7.6	11.8	12.3	10.2	13.1
D	3.4	5.9	4.3	6.2	6.3	ND

Because the measurements within each station are probably correlated, the rankings are done within each station. This test is very similar to Friedman's two-way analysis of variance by ranks, with the added stipulation that we are looking for a monotonic increase (or decrease) through time. Under the null hypothesis of no trend, one expects the ranks 1–6 to appear in random order for each station. Under the alternative hypothesis of increasing trend, one expects that the ranks would increase also. (If the alternative hypothesis is one of a decreasing trend through time, then the year and the data would be arranged in reverse order.)

Ranking the measurements within each station, and then summing the ranks for each year yields the following (note that there is one instance of tied values and one instance of a non-detect):

	YEAR					
	1	2	3	4	5	6
STATION						
A	3	2	1	6	4	5
B	1	2	3	4	5.5	5.5
C	2	1	4	5	3	6
D	2	4	3	5	6	1
R_j	8	9	11	20	18.5	17.5

where R_j is the sum of the ranks for the jth year.

Page's test is based upon the test statistic

$$L = \sum_{j=1}^{k} jR_j = R_1 + 2R_2 + 3R_3 + \ldots + kR_k \tag{7.51}$$

which, for these data, is

$$L = 1 \cdot 8 + 2 \cdot 9 + 3 \cdot 11 + 4 \cdot 20 + 5 \cdot 18.5 + 6 \cdot 17.5 = 336.5 \qquad (7.52)$$

statistically significant at the 0.01 level (the table used here is from Siegel and Castellan 1988, Table N). If we did not have any information prior to looking at the data in which direction the trend might go, then to do a two-sided test, the final P-value obtained from the table would need to be doubled to 0.02.

7.11 MORE ON TEMPORAL CORRELATION

When correlation among sequential observations is present, and if a sufficiently long time sequence of equally spaced observations is available (at least 30 or more observations), then the statistical modelling approach known as time series analysis may be applied. The accepted classical reference for time series analysis is the book by Box and Jenkins (1976); Diggle (1990) has a biostatistical orientation rather than an econometric one. The actual modelling process for time series analysis is quite complex; in this and the following sections, we emphasize the concept of data dependence through serial correlation and how to account for it in some basic parametric and non-parametric tests.

Like many statistical models, a time series model consists of two parts: a systematic or deterministic component, changing over time in a regular and predictable way, and a random component (Montgomery and Reckhow 1984). The systematic component can arise from the effects of a trend, periodicity (e.g. seasonal effects), or serial correlation (i.e. the observation at time t is correlated with previous observations, even when trend and seasonality have been accounted for).

The presence of serial correlation definitely affects some of the most commonly used statistical measures, most notably that of variances and their resulting standard errors. To start with a simple model, suppose there is correlation between adjacent observations, even after trends have been removed from the data. The natural estimate for the overall series mean would be $\bar{Y}$, the average of the n observations over time. If these observations were independent (which they are not), the variance associated with $\bar{Y}$ would be σ^2/n. Now, with serial correlation present,

$$\mathrm{Var}(\bar{Y}) = \frac{\sigma^2}{n} \cdot \left[1 + 2\left(\frac{n-1}{n}\right) r_1 \right] \qquad (7.53)$$

where r_1 represents the serial correlation between adjacent observations (Bayley and Hammersley 1946).

Table 7.4 (from Conquest 1993) shows by how much the familiar standard error for $\bar{Y}$, $\sigma/\sqrt{n}$, becomes inflated by the presence of such serial correlation. For a given number n, the multiplier changes from about 1.1 to about 1.7. Hence even if r_1 is known only approximately, one can get an idea of how much

Table 7.4 Multiplier of $(s\sqrt{n})$ for $Var(\bar{Y})$ when first-order autocorrelation is present (from Conquest, 1993). Reproduced with permission of Kluwer Academic Publishers

n	r_1								
	0.1	0.2	0.3	0.4	0.5	0.6	0.7	0.8	0.9
5	1.08	1.15	1.22	1.28	1.34	1.40	1.46	1.51	1.56
6	1.08	1.15	1.22	1.29	1.35	1.41	1.47	1.53	1.58
7	1.08	1.16	1.23	1.30	1.36	1.42	1.48	1.54	1.59
8	1.08	1.16	1.23	1.30	1.37	1.43	1.49	1.55	1.60
9	1.09	1.16	1.24	1.31	1.37	1.44	1.50	1.56	1.61
10	1.09	1.17	1.24	1.31	1.38	1.44	1.50	1.56	1.62
15	1.09	1.17	1.25	1.32	1.39	1.46	1.52	1.58	1.64
20	1.09	1.17	1.25	1.33	1.40	1.46	1.53	1.59	1.65
25	1.09	1.18	1.26	1.33	1.40	1.47	1.53	1.59	1.65
30	1.09	1.18	1.26	1.33	1.40	1.47	1.53	1.60	1.66
∞	1.10	1.18	1.26	1.34	1.41	1.48	1.53	1.61	1.67

larger the standard error for $\bar{Y}$ becomes due to the presence of autocorrelation. For a regression model where $Y(t)$ is also considered to be a linear function of various predictor variables, Cochrane and Orcutt (1949) have derived results for doing least squares regression in the presence of autocorrelated error terms.

We can go through the same type of procedure for computing the required sample size to achieve a 95% confidence interval of a desired length to see how the required sample size is affected by serial correlation. A generalization of a well-known formula for sample size computations based on such a confidence interval approach leads to the following approximation (Gilbert 1987).

$$n = Z^2_{\alpha(2)} \cdot \left(\frac{\sigma^2}{d^2}\right) \cdot (1 + 2r_1) \qquad (7.54)$$

where d is the maximum amount by which the sample mean is 'allowed' to deviate from the true mean, and $Z_{\alpha(2)}$ represents the two-sided cut-off value from a standard normal distribution that leads to a $(1 - \alpha) \cdot 100\%$ confidence interval. As an example, if we wanted a 95% confidence interval to be within 20 units of the true mean, with a standard deviation of 50 and in the presence of serial correlation of 0.30, the estimated sample size would be 38.4, or 39 sampling times.

The above equation may also be re-expressed as the following:

$$n = Z^2_{\alpha(2)} \cdot \left(\frac{1}{d^2/\sigma^2}\right) \cdot (1 + 2r_1) \qquad (7.55)$$

Because the relationship between d^2 and σ^2 in the equation is a ratio, as long as we can specify the maximum allowable error in terms of standard deviations, it is not necessary to specify each quantity separately. The above example could

Table 7.5 Approximate sample size for 95% confidence intervals, as a function of maximum allowable error in standard deviations, and serial correlation

d/σ:	0.2	0.3	0.4	0.5
r_1				
0.0	96	43	24	15
0.1	115	51	29	18
0.2	134	60	34	22
0.3	157	68	38	25
0.4	173	77	43	28
0.5	192	85	48	31
0.6	211	94	53	34
0.7	230	102	58	37
0.8	250	111	62	40
0.9	269	112	67	43

be restated by saying that the maximum allowable error should be two-fifths of a standard deviation.

This idea may prove useful when no prior estimate of the standard deviation is available. Table 7.5 gives approximate required sample sizes (numbers rounded to the nearest integer) for 95% confidence intervals ($Z = 1.96$), various maximum allowable errors (stated in terms of standard deviations), and various values of r_1.

Tables 7.4 and 7.5 illustrate how serial correlation, by reducing the amount of new information from an additional data point, can result in a substantial increase in the number of required sampling points in order to achieve the same objective.

Any test that relies upon the independence of the observations is affected by the presence of serial correlation; as an example, consider the standard two-sample t-test, Suppose one wants to consider whether a shift at some point in a time series is statistically significant; that is, comparing the value of $\bar{Y}_B$ to $\bar{Y}_A$, where the two means are comprised of points before and after some known intervention. To assess the significance of the observed difference $(\bar{Y}_B - \bar{Y}_A)$ it is necessary to consider the magnitude of the difference with respect to its variance:

$$\text{Var}(\bar{Y}_B - \bar{Y}_A) = \text{Var}(\bar{Y}_B) + \text{Var}(\bar{Y}_A) - 2\,\text{covariance}\,(\bar{Y}_B, \bar{Y}_A) \qquad (7.56)$$

For independent data points, the covariance term is zero, but this will not be the case when autocorrelation is present for adjacent observations. The formula for the variance of each mean in the presence of autocorrelation has been stated above. The covariance term is as follows:

$$\text{covariance}\,(\bar{Y}_B, \bar{Y}_A) = \frac{\sigma^2}{n^2} \cdot r_1 \qquad (7.57)$$

so that the variance for the difference between the two means, instead of the usual $2 \cdot \sigma^2/n$, becomes

$$\mathrm{Var}(\bar{Y}_\mathrm{B} - \bar{Y}_\mathrm{A}) = \frac{2 \cdot \sigma^2}{n} \cdot \left[1 + \frac{(2n - 3)}{n} \cdot r_1\right] \qquad (7.58)$$

When r_1 = zero, this expression simplifies to $2 \cdot \sigma^2/n$, as expected. For large n, the extra multiplier on the right is approximately equal to $(1 + 2 \cdot r_1)$, values for which may be seen in the last row of Table 7.4.

7.12 NON-PARAMETRIC TESTS FOR STEP CHANGE OR TREND FOR SERIALLY CORRELATED DATA

When data are serially correlated, one can no longer proceed with the usual battery of statistical tests based upon mutually independent observations. For two widely used non-parametric tests, the Mann–Whitney–Wilcoxon (MWW) two-sample rank test (to detect the presence of a step change) and the test on Spearman's r_s (to detect the presence of monotonic trend), Lettenmaier (1976, 1977) and Montgomery and Reckhow (1984) have proposed modifications to each of these tests, where the data dependence occurs in the form of lag 1 serial correlation (i.e. correlation between data points that are adjacent in time; this implies that the sampling over time has occurred at equally spaced intervals). What these authors did was to propose modified critical values for the MWW test for step change and the Spearman r_s test for association, for various sample sizes. Table 7.6, based on algorithms in Montgomery and Reckhow (1984), shows modified Spearman r_s values for sample sizes n = 30, 50 and 100 for a two-sided test at the 0.05 level of significance. Not surprisingly, the critical value for $|r_s|$ increases as serial correlation increases. Table 7.6 also displays the approximate corresponding values of n associated with the new critical $|r_s|$ values if one were dealing with independent data points. So, for example, a data set of 30 points taken through time, with a serial correlation of 0.30 (this is the correlation that remains between adjacent points even after trends have been removed), yields the same amount of information as an independent set of 17 data points.

Table 7.7 shows modified lower and upper critical values for the MWW two-sample test for changes in location. The form of the test statistic used here is

$$U = \sum_{i=1}^{n_1} R(X_i) - n_1 \frac{(n_1 + 1)}{2} \qquad (7.59)$$

where n_1 is the sample size for the smaller sample. Table 7.7 is also based upon algorithms and correction factors for sample sizes n = 30, 50, and 100 for a two-sided test at the 0.05 level of significance. As expected, the higher the level of serial correlation, the more extreme the value of the test statistic has to be in order to declare a statistically significant difference. For example, suppose there were 10 equally spaced time points before an intervention event and 20 after it.

Table 7.6 Modified absolute values for statistical significance of Spearman r_s (two-sided 0.05 level) as a function of serial correlation. Corresponding sample sizes for independent data in parentheses

Serial correlation	$n = 30$	$n = 50$	$n = 100$
0.0	0.362	0.279	0.197
	(30)	(50)	(100)
0.1	0.407	0.330	0.234
	(24)	(36)	(71)
0.2	0.455	0.355	0.244
	(19)	(31)	(65)
0.3	0.486	0.368	0.273
	(17)	(28)	(52)
0.4	0.535	0.423	0.305
	(14)	(22)	(42)
0.5	0.592	0.487	0.346
	(12)	(17)	(33)
0.6	0.656	0.541	0.393
	(10)	(14)	(26)
0.7	0.733	0.634	0.463
	(08)	(10)	(19)

(Note that the MWW test itself does not demand that the time points be equally spaced; it is the notion of serial correlation that implies equal spacing.) With no temporal correlation, the critical values for the test statistic are 55 and 145. Serial correlation of 0.2 will widen these values to 48 and 152; serial correlation of 0.5 will further widen the values to 31 and 169.

When testing for the significance of a step change (whether one is using a parametric or non-parametric approach), it is important to keep in mind whether the 'cut point' between the two samples has been chosen on the basis of prior knowledge, or only after having looked at the data. If the cut point has been chosen on the basis of prior knowledge, such as knowing when a particular intervention event occurred, then the cut point is not based upon seeing the data. If, however, the cut point has been chosen based upon looking at the data to decide where is the most likely point for the step change to have occurred, then we must realize that we are looking at a test statistic based upon the maximum difference between the two time periods, and that was based upon the data-driven choice of the cut point. In such a case, the test statistic must be modified to take this phenomenon into account; this can be a fairly complex undertaking and is beyond the scope of this chapter. The reader is referred to Ratcliffe (1970), which includes a detailed appendix for modifying the two-sample t-test when comparing eggshell strength between an early and late period whose cut point is based upon the data.

Table 7.7 Modified lower and upper critical values for the Mann–Whitney–Wilcoxon statistic, as a function of lag 1 serial correlation 0

$n = 30$ n_1, n_2	$r_1 = 0_1$	$r_2 = 0.2_1$	$r_1 = 0.4_1$	$r_1 = 0.6_1$
5, 25	27, 98	24, 101	16, 109	12, 113
6, 24	33, 111	29, 115	20, 124	15, 129
7, 23	40, 121	35, 126	24, 137	18, 143
8, 22	45, 131	39, 137	27, 149	21, 155
9, 21	50, 139	44, 145	30, 159	23, 166
10, 20	55, 145	48, 152	33, 167	25, 175
11, 19	58, 151	51, 158	35, 174	27, 182
12, 18	61, 155	53, 163	37, 179	28, 188
13, 17	63, 158	55, 166	38, 183	29, 192
14, 16	64, 160	56, 168	39, 185	29, 195
15, 15	64, 161	56, 169	39, 186	29, 196
$n = 50$				
10, 40	119, 281	102, 298	84, 316	61, 339
11, 39	130, 299	111, 318	92, 337	66, 363
12, 38	141, 315	120, 336	100, 356	72, 384
13, 37	151, 330	129, 352	107, 374	77, 404
14, 36	161, 343	137, 367	114, 390	82, 422
15, 35	169, 356	144, 381	119, 406	86, 439
16, 34	177, 367	151, 393	125, 419	90, 454
17, 33	184, 377	157, 404	130, 431	94, 467
18, 32	190, 386	162, 414	134, 442	97, 479
19, 31	196, 393	167, 422	138, 451	100, 489
20, 30	200, 400	171, 429	141, 459	102, 498
$n = 100$				
10, 90	279, 621	251, 649	208, 692	167, 733
15, 85	434, 841	390, 885	323, 952	259, 1016
20, 80	573, 1027	514, 1086	425, 1175	342, 1258
25, 75	691, 1184	620, 1255	514, 1361	413, 1462
30. 70	789, 1311	708, 1392	587, 1513	471, 1629
35, 65	866, 1409	777, 1498	644, 1631	517, 1758
40, 60	921, 1479	827, 1573	685, 1715	550, 1850
45, 55	955, 1520	856, 1619	709, 1766	570, 1905
50, 50	966, 1534	866, 1634	718, 1782	577, 1923

7.13 SUMMARY

This chapter has focused on methods associated with non-parametric measures of association. We have investigated non-parametric tests for trend, either based upon ranks (Spearman's r_s) or on the counting concordant and discordant pairs (Kendall's au). The advantages of taking distribution-free approaches, apart from freeing the user from the assumption of the normal distribution, is that these basic tests extend to data taken in different seasons and at different stations. Non-parametric slope estimation is less likely to be influenced by wildly behaving data points. The tests can be modified to handle non-detects and missing values, and they do not depend upon equally spaced sampling in the predictor variable, such as time. Serial correlation in a time sequence of data introduces redundancy of information in the data and reduces the effective size of the sample. Time series modelling and analysis is a fairly complex subject. This chapter has introduced the reader to some of the basic concepts, and we have concluded with modifications of the traditional MWW test and Spearman's r_s test in the presence of serial correlation.

REFERENCES

Alvo M and Cabilio P (1993) Tables of critical values of rank tests for trend when the data is incomplete. Technical Report No. 230, Laboratory for Research in Statistics and Probability, Room 611 Dunton Tower, Carleton University, Ottawa, Ontario K1S 5B6 Canada

Alvo M and Cabilio P (1994) Rank test of trend when data are incomplete. *Environmetrics*, **5**, 21–27.

Alvo M and Cabilio P (1995) Rank correlation methods for missing data. *The Canadian Journal of Statistics*, **23**, 345–358.

Anscombe FJ (1973) Graphs in statistical analysis. *The American Statistician*, **27**, 17–21.

Bayley GV and Hammersley JM (1946) The effective number of independent observations in an autocorrelated time series. *Journal of the Royal Statistical Society* (supplement), Series B, **8**, 184–197.

Box GEP and Jenkins GM (1976) *Time Series Analysis: Forecasting and Control*, 3rd edn, Prentice-Hall, Englewood Cliffs NJ, 598 pp.

Cleveland WS (1979) Robust locally weighted regression and smoothing scatterplots. *Journal of the American Statistical Association*, **74**, 829–836.

Cleveland WS (1981) LOWESS: A program for smoothing scatterplots by robust locally weighted regression. *The American Statistician*, **35**, 54.

Cochrane D and Orcutt GL (1949) Application of least-squares regression to relationships containing autocorrelated error terms. *Journal of the American Statistical Association*, **71**, 961–967.

Conquest, LL (1993) Statistical approaches to environmental monitoring: did we teach the wrong things? *Environmental Monitoring and Assessment*, **26**, 107–124.

Diggle, PJ (1990) *Time Series: A biostatistical Introduction*, Clarendon Press, Oxford, 257 pp.

Durbin J and Watson GS (1951) Testing for serial correlation in least squares regression, I and II, *Biometrika*, **37**, 409–428 and **38**, 159–178.

Furness R and Hutton M (1979) Pollutant levels in the Great Skua *Cathartica skua. Environmental Pollution*, **19**, 261–268.

Gilbert RO (1987) *Statistical Methods for Environmental Pollution Monitoring*. Van Nostrand Reinhold, New York, p 205.

Helsel DR and Hirsch RM (1992) *Statistical Methods in Water Resources*. Elsevier, New York, 522 pp.

Hirsch RM and Slack JR (1984) A nonparametric trend test for seasonal data with serial dependence. *Water Resources Research*, **20**, 727–732.

Hirsch RM, Slack JR, and Smith RA (1982) Techniques of trend analysis for monthly water quality data. *Water Resources Research*, **18**, 107–121.

Hollander M and Wolfe DA (1973) *Nonparametric Statistical Methods*. Wiley, New York. 503 pp.

Hughes, J and Millard SP (1988) A tau-like test for trend in the presence of multiple censoring points. *Water Resources Bulletin*, **24**, 521–531.

Kaarsemaker L and van Wijngaarden A (1953) Tables for use in rank correlation. *Statistica Neerlander*, **7**, 41–54.

Kendall MG (1975) *Rank Correlation Methods*, 4th edn. Charles Griffin, London, 202 pp.

Lettenmaier DP (1976) Detection of trends in water quality data from records with dependent observations. *Water Resources Research*, **12**, 1037–1046.

Lettenmaier DP (1977) *Detection of Trends in Stream Quality Monitoring Network Design and Data Analysis*. Technical Report No. 51, CW Harris Hydraulic Laboratory, Department of Civil Engineering, University of Washington, Seattle WA 98195, USA.

Mann BH (1945) Non-parametric tests against trend. *Econometrica*, **13**, 245–259.

Montgomery RH and Reckhow KH (1984) Techniques for detecting trends in lake water quality. *Water Resources Bulletin*, **20**, 43–52.

Newton I and Haas MB (1984) The return of the sparrowhawk. *British Birds*, **77**, 47–70.

Page EB (1963) Ordered hypotheses for multiple treatments: a significance test for linear ranks. *Journal of the American Statistical Association*, **58**, 216–230.

Ratcliffe DA (1970) Changes attributable to pesticides in egg breakage frequency and eggshell thickness in some British birds. *Journal of Applied Ecology*, **7**, 67–115.

Rohlf FJ and Sokal RR (1995) *Statistical Tables*, 3rd edn. WH Freeman, NY, 199 pp.

Sen PK (1968) Estimates of the regression coefficient based on Kendall's tau. *Journal of the American Statistical Association*, **63**, 1379–1389.

Shelly AA (1994) Statistical power for detecting trends in monitoring programs using Correlation. MS. thesis, University of Washington, c/o Quant. Ecology & Resource Management Program, Box 351720, Seattle WA 98195, USA, 80 pp.

Siegel S and NJ Castellan (1988) *Nonparametric Statistics for the Behavioral Sciences*, 2nd edn. McGraw-Hill, NY, 399 pp.

Smith RA, Hirsch RM and Slack JR (1982) *A Study of Trends in Total Phosphorus Measurements at NASQAN Stations*. US Geological Survey Water-Supply Paper 2190, US Geological Survey, Alexandria VA, USA.

Sokal RR and Rohlf FJ (1995) *Biometry, The Principles and Practices of Statistics in Biological Research*, 3rd edn. WH Freeman, NY, pp. 394–396.

Sparks TH and Rothery P (1996) Resampling methods of ecotoxicological data. *Ecotoxicology*, **5**, 197–207.

Theil H (1950) A rank-invariant method of linear and polynomial regression analysis. Part 3, in *Proceedings of Koninalijke Nederlandse Akademie van Wetenschatpen* A, **53**, 1397–1412.

Van Belle G and Hughes J (1984) Non parametric tests for trend in water quality. *Water Resources Research*, **20**, 127–136.

Weisberg, S. (1985) *Applied Linear Regression*, 2nd edn. Wiley, NY, 324 pp.

Organochlorines in Bird of Prey Eggs: a Terrestrial Case Study

TIM SPARKS, IAN NEWTON, LOIS DALE AND DAN OSBORN

Institute of Terrestrial Ecology, Monks Wood, UK

8.1 INTRODUCTION

Organochlorine pesticides (OCs) have been implicated in a range of wildlife poisoning incidents (e.g. see Sheail 1986); ultimately leading to the withdrawal of DDT and dieldrin in the 1970s and 1980s. Certain species towards the top of the terrestrial food chains are potentially most at risk from these chemicals because of bioaccumulation in body tissues. Several species of birds of prey are thought to have declined in numbers as a result of contamination by OCs. Decline was perceived as occurring through two channels: lethal effects on adult birds and reduced reproduction associated with eggshell thinning. Population reductions were more marked in the most arable areas of Britain, in the east, where pesticide use was greater.

Since 1963, staff of the Institute of Terrestrial Ecology Monks Wood, and its predecessor the Nature Conservancy (Council), have been monitoring contamination levels in the tissues and eggs of birds of prey (Cooke *et al.* 1982; Newton *et al.* 1993). These species, near the tops of food chains, make ideal subjects for studying the bioaccumulation of toxic substances.

In this scheme, tissue monitoring relies on opportunistic sampling, as recovered bodies of dead birds are sent to Monks Wood for chemical analysis. Carcasses of six species are requested to represent predatory birds reliant on terrestrial and aquatic prey: sparrowhawk *Accipiter nisus*, kestrel *Falco tinnunculus*, barn owl *Tyto alba*, heron *Ardea cinerea*, kingfisher *Alcedo atthis* and great-crested grebe *Podiceps cristatus*. It would be inappropriate and undesirable to cull these animals because they are of conservation interest. This necessitates reliance on encounter sampling. All carcasses are requested, regardless of suspected cause of death: some arise from poisoning incidents, others from natural causes or accidents. Levels of OCs in tissue residues vary between species, partly reflecting differences in diet, and, within species, tend

---

Statistics in Ecotoxicology. Edited by T. Sparks. © 2000 John Wiley & Sons Ltd

to be higher in arable areas (Newton *et al.* 1993). In general, over the years, there has been a trend downwards in DDT and dieldrin residues. The pattern has been less clear for PCBs which are largely non-agricultural in origin. Mercury residues are also monitored and have declined. These patterns have been associated with a return of species to areas abandoned during population declines (Newton and Haas 1984; Ratcliffe 1993).

Sampling for eggs, other than for heron, is also opportunistic. Reliance is placed on obtaining samples from deserted nests or unhatched (addled) eggs. There has to be some debate on whether these eggs are representative of the population as a whole. Two important points to make here are that, first, random sampling of eggs from a declining bird population would not be acceptable from a conservation viewpoint and, secondly, that the method of collection has been consistent over the years and as such allows an examination of change in residues of eggs collected in similar circumstances. Eggshell thinning has been associated with reduced brood size in sparrowhawk (Newton *et al.* 1986). Eggshell thickness seems to be improving in sparrowhawk (Newton 1986), merlin *Falco columbarius* (Newton *et al.* in press), peregrine *Falco peregrinus* (Newton *et al.* 1989; Ratcliffe 1980) and golden eagle *Aquila chrysaetos* (Newton and Galbraith 1991).

8.2 ANALYSING DATA FROM PEREGRINE EGGS

Data on the levels of contaminants found in animal tissue present a number of statistical challenges. Data presented in the past have often been analysed by straightforward study of means and measures of variation. However, as the size of the data sets increases it becomes important to analyse the data carefully to avoid bias in interpretation or unreasonable conclusions, and to extract information about temporal or spatial pattern. Careful statistical analysis is particularly important for large data sets collected over a long period of time during which a number of factors can influence the data. For this case study, we have chosen a large data set collected over the past 35 years. It concerns the eggs of the peregrine, a species feeding mainly on larger birds than does the sparrowhawk. Its main prey over much of Britain is the feral pigeon, but it eats a wide range of other birds, including seabirds.

The peregrine suffered a marked population decline in Britain from which it has now largely recovered (Ratcliffe 1980; Crick and Ratcliffe 1995). A single egg record dates back to 1961, otherwise monitoring proper began in 1963. Chemical analysis was split initially between the Laboratory of the Government Chemist, Glasgow University Veterinary School and Monks Wood Research Station with cross-checks between them. More recently chemical analysis has only been done at Monks Wood. Each egg was analysed for DDE (a metabolite of DDT), HEOD (a component of dieldrin and a metabolite of aldrin), total PCB (organochlorine chemicals derived from various industrial products; post-1966 data only) and, largely since 1986, mercury. All data for contaminants are in parts per million wet weight. The three OCs were subject to voluntary withdrawal resulting in progressive reduction in use from the 1960s, until their final withdrawal in 1986. PCBs were no longer used in likely polluting situations after 1971, but these chemicals continue to escape to the environment from products made in earlier years. Mercury exists in the environment naturally, but is greatly elevated by human activities. Some peregrine eggs are from the same clutch, where eggs tend to be more similar to one another than from different clutches (Newton *et al.* 1989). Because of the low incidence of this feature in the data set this feature has been ignored for the purpose of this case study. Shell index was calculated as weight(mg)/(length(mm) $\times$ breadth(mm)). The pre-DDT shell index of peregrine eggs had a mean value of 1.82 (Ratcliffe 1980).

Between 1961 and 1996, results from 645 eggs are available. While records are complete for DDE, there are 644 records for HEOD (1 missing), 612 for PCB (33 missing) and only 239 for mercury, because of its later incorporation into the monitoring scheme. Mercury determination is, however, missing from only four eggs in the 1986–96 period. 518 records of shell index are available. The distribution of egg numbers between years is uneven, as shown in Table 8.1.

Chemical analysis of these contaminants is subject to a detection limit; for example 0.1 ppm in HEOD and DDE. With monitoring schemes such as these, instrumental improvements often result in the lowering of a detection limit over time. In the statistical analysis of data, numerical values are desirable for those below detection limit (BDL), or non-detect, values. The allocation of numerical values to BDL records is an unresolved subject area and there is insufficient space here to go into too much detail. Allocating either detection limit or half detection limit to those values may be unsatisfactory if the detection limit changes over time. An alternative may be to honour the highest

Table 8.1 Number of peregrine eggs analysed for each of the years 1961–96 and summarized by decade and overall

Year	No. of eggs	Year	No. of eggs	Year	No. of eggs	Year	No. of eggs
		1970	7	1980	17	1990	2
1961	1	1971	19	1981	43	1991	48
1962	0	1972	8	1982	35	1992	23
1963	10	1973	12	1983	17	1993	12
1964	3	1974	11	1984	37	1994	27
1965	6	1975	28	1985	29	1995	14
1966	8	1976	17	1986	28	1996	13
1967	12	1977	21	1987	20		
1968	13	1978	19	1988	25		
1969	13	1979	26	1989	21		
Decade total	66		168		272		139
						Overall total	645

detection limit during the period of study, relabelling all data below this as BDL. Here we have replaced BDL values by zero, as the purpose of this case study is to illustrate the types of statistical analysis that can be done on data sets with long temporal and wide spatial scales.

8.3 EXPLORATORY DATA ANALYSIS

Exploratory data analysis can be of great value in understanding the nature of the data we are dealing with. This practice becomes increasingly important as data sets get larger and harder to comprehend. Histograms of the raw variables suggest that the residue data are far removed from those expected from a normal distribution, whilst shell index might fit that description much more easily (Figure 8.1).

In other words, these raw data, with the exception of shell index, are highly right skewed. This is typical of many chemical determinations in wildlife samples, highly influenced by a few super-elevated values. Such skew distributions often obscure relationships or confound interpretation of relationships between variables. A matrix scatterplot of the data suggests that patterns of relationships are not obvious (Figure 8.2). A matrix of correlation coefficients is presented in Table 8.2. Because of the large sample size, all of these are significant, except for the HEOD/PCB and Hg/SI relationships. However, standard significance levels are likely to be misleading because of the skewed nature of the data, and significance levels via randomization procedures (Sparks and Rothery 1996) or rank correlation, as shown in the second part of Table 8.2, are more reliable. This reveals a stronger pattern of relationship, as the influence of extreme

Table 8.2 Correlations and Spearman rank correlations between the recorded variables. Those in bold are nominally significant, at least at the 5% level, but note the caution in the text

Correlations (Pearson)

	HEOD	DDE	PCB	HG
DDE	**0.257**			
PCB	0.060	**0.254**		
HG	**0.186**	**0.376**	**0.457**	
SI	**−0.110**	**−0.370**	**−0.173**	−0.137

Spearman rank correlations

	HEOD	DDE	PCB	HG
DDE	**0.635**			
PCB	**0.441**	**0.555**		
HG	**0.280**	**0.444**	**0.333**	
SI	**−0.251**	**−0.441**	**−0.228**	−0.091

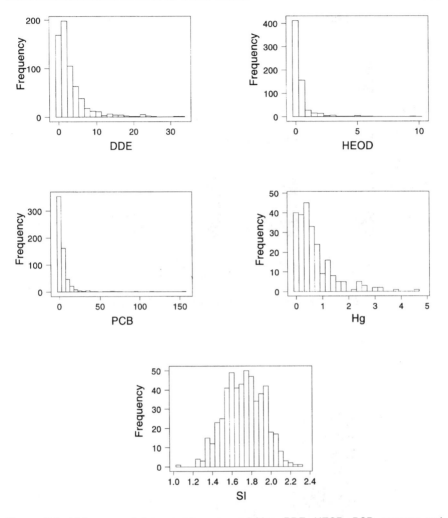

Figure 8.1 Histograms of the peregrine egg variables: DDE, HEOD, PCB, mercury and shell index. Extreme high values for HEOD, DDE, PCB and Hg are 9.4, 33, 154 and 4.6 ppm respectively.

points is downweighted. Once again the Hg/SI relationship is not significant. Because rank correlations measure monotonic (a consistent rise or fall), not just linear, relationships they may have higher statistical power in identifying trends.

Many studies on eggshell thinning have associated thinning with high DDE residues in eggs. Figure 8.3 shows this relationship in a little more detail. Added to this scatterplot is a smoothed line (using LOWESS – locally weighted scatter-plot smoother) which should reveal the underlying pattern without any preconception of what that relationship is.

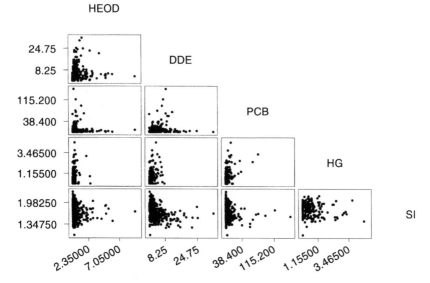

Figure 8.2   A matrix scatterplot of the recorded variables.

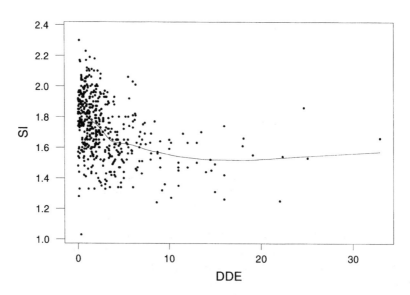

Figure 8.3   A scatterplot of the relationship between shell index and untransformed DDE residues.

The LOWESS line in this scatter suggests that there may be an underlying relationship; increasing DDE up to about 8 ppm associated with lowered shell index values, after which increased DDE seems to have little effect. This relationship is hard to see in a plot of the raw data, being confused by the high numbers of eggs with low DDE values and small numbers with high DDE values. The fitted shell index value of this LOWESS line for DDE equal to zero is 1.83, very close to the pre-DDT level of 1.82 quoted above. In light of the suggested pattern, it is not surprising that a rank correlation produces a larger correlation coefficient.

The relationship may be clarified by transforming the DDE values, and with data such as these which are highly skew a log transformation may be appropriate. Figure 8.4 displays the relationship between shell index and $\log(x + 1)$ transformed DDE values. A LOWESS line has been added to this. Several features are more apparent in this scatterplot. The relationship between shell index and log transformed DDE values looks more nearly linear; variability about the LOWESS lines looks more uniform but very low shell index values are still evident at low residue levels of DDE. Once again, the 'intercept' of the LOWESS line (i.e. at DDE zero level) is 1.83.

Figure 8.5 further enhances our interpretation of this plot by labelling records from different decades (1961–69, 1970–79, 1980–89, 1990–96) with different symbols. This suggests that there may not be equal levels of DDE in the four decades, higher levels in the 1960s and lower ones in more recent decades, a feature which we will examine in more detail later.

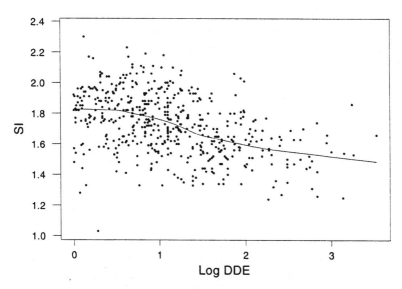

Figure 8.4 A scatterplot showing the relationship between shell index and $\log(x + 1)$ transformed DDE values. A LOWESS line is superimposed.

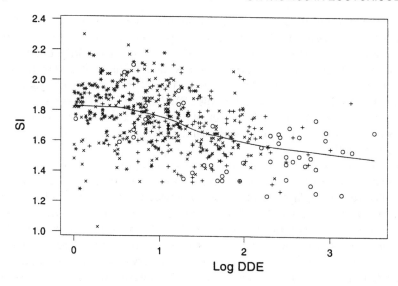

Figure 8.5 The scatterplot of shell index against $\log(x + 1)$ transformed DDE values. LOWESS line superimposed. Decadal symbols are $\bigcirc$(1960s), + (1970s), × (1980s) * (1990s).

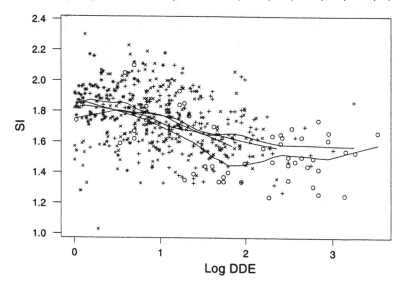

Figure 8.6 A repeat of the previous figure but with separate LOWESS lines for each decade.

Figure 8.5 suggests temporal changes in DDE levels, so we need also to examine if the shell index/DDE relationship changed over the years. Figure 8.6 repeats this figure but with separate LOWESS lines for each decade. In this example the four decadal LOWESS lines can be considered to be broadly similar, suggesting that the relationship between DDE and shell index has not changed over time. In other words, there is no indication that the peregrines have developed any resistance to DDE.

In this section we have examined the form of the data, found some broad relationships between variables and have detected some evidence of trends in time. Now we turn our attention to looking at patterns through time.

8.4　TEMPORAL PATTERNS

Figure 8.7 displays a plot of raw DDE data against year of collection. The wide scatter makes any underlying trend hard to interpret, although using a LOWESS line in the lower part of the figure does help, suggesting a non-linear decline over time.

As we have already seen, interpretation may be improved by transforming the variable. Figure 8.8 repeats Figure 8.7, but with DDE values $\log(x + 1)$ transformed. The relationship looks much more nearly linear on a log scale and this is further confirmed by the use of a LOWESS line.

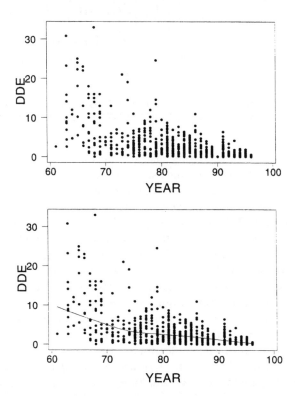

Figure 8.7 A scatter of raw DDE data against year of collection (after 1900), the lower graph also includes a LOWESS line to aid interpretation.

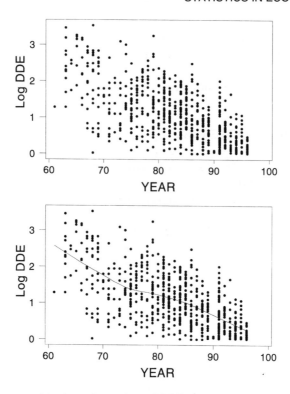

Figure 8.8 A scatter of log(*x* + 1) transformed DDE data against year of collection (after 1900), the lower graph also includes a LOWESS line to aid interpretation.

Figure 8.9 repeats these graphs against time for the other four variables. A reduction in HEOD levels is suggested by the LOWESS line, although an inspection of the bare scatter would be less convincing. The pattern for PCB is ambiguous, that for mercury heavily distorted by a few early records, and clear improvements in shell index are suggested, mirroring the decline in DDE, although without the LOWESS line the trend is not so apparent.

There may be some justification for working with annual mean concentrations rather than with individual data points. This serves two purposes. First one might expect eggs within a year to be more similar to one another than eggs from different years, so there may be some problems in considering eggs from the same year as 'independent' in a study of trends through time. Secondly, we would expect, from the central limit theorem (see Chapter 1), that means would be more normally distributed than the raw data. One thing for sure is that reducing the data to a series of annual means does aid interpretation. Data of this type have traditionally been presented as geometric means rather than as arithmetic means. Figure 8.10 displays these data as both arithmetic and geometric means. The two patterns are broadly similar; HEOD

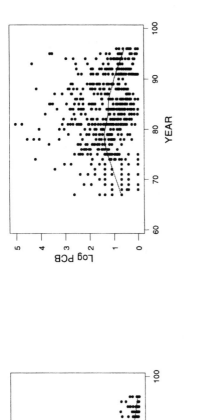
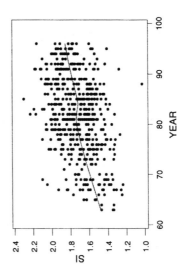
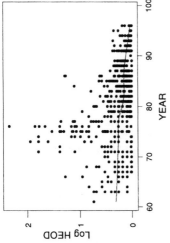
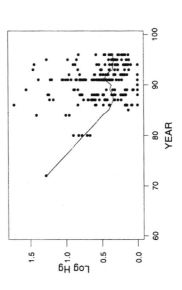

Figure 8.9 Scatterplots of log(x + 1) transformed HEOD, PCB, mercury and untransformed shell index against year of sampling (after 1900). LOWESS lines added.

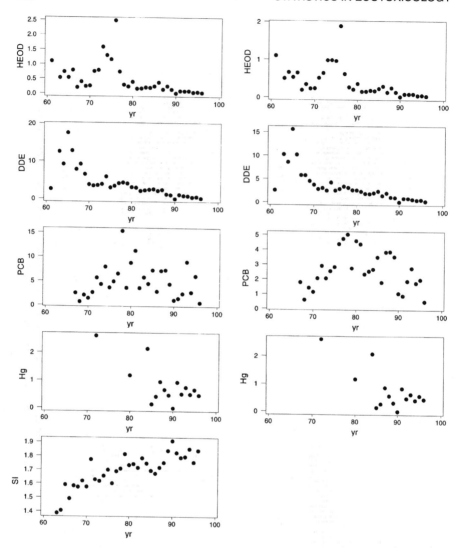

Figure 8.10 A plot of arithmetic (left) and geometric (right) means of the variables plotted against time (year after 1900). No geometric mean was considered necessary for shell index.

plots suggest a decline over time but with a peak in the 1970s, the decline is more pronounced for DDE although the lone 1961 record is influential, the geometric means for PCB more convincingly suggest a peak in the 1980s, the mercury pattern is again distorted by a few pre-1986 eggs and shell index shows convincing improvements over time.

These scatterplots, based on annual mean values, weight the result from each year equally, irrespective of the numbers of eggs collected in that year. Because

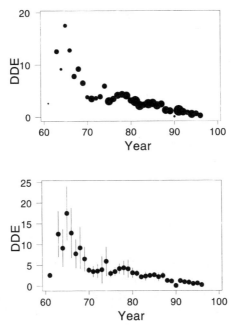

Figure 8.11   An influence plot of arithmetic mean DDE against year of collection (after 1900), symbol size proportional to sample size (upper) and with two sigma limits (two standard errors) (lower).

egg collection is largely opportunistic, numbers vary between years, as shown in Table 8.1. So, the 1961 single egg is over-influential is this plot, while earlier plots are more influenced by years with larger egg numbers. Which is appropriate, annual means or all raw data? Before we attempt to answer this question, we look at two further ways of presenting the DDE arithmetic means. The upper part of Figure 8.11 shows the data presented as an influence plot with symbol size related to sample size; 1961 barely shows up on this scale. The lower part of the figure shows means with two standard error limits; much more variable in the early part of the series, but no limits showing for 1961 since variability cannot be estimated from a single observation.

If we are to do a regression of variable on year to examine for trends, there is some suggestion that an element of pseudoreplication is involved. This is a debatable point since the only link between the eggs is that they are collected in the same year. However, one might expect them to be more similar than eggs from different years.

Table 8.3 summarizes the regression of $\log(x + 1)$ transformed DDE value on year, based on all eggs. A strong relationship is apparent. The parameters of this model suggest that, given no further environmental input, mean DDE levels might return to zero (more accurately to non-detect levels) in the year 2003, although extrapolation beyond range should always be undertaken with

Table 8.3 Regression of log (*x* + 1) DDE on year. All data points included

---

Regression analysis

```
The regression equation is
logdde = 5.5 - 0.0540 YEAR
```

Predictor	Coef	StDev	T	P
Constant	5.5481	0.2276	24.38	0.000
YEAR	-0.053987	0.002763	-19.54	0.000

S = 0.5866   R—Sq = 37.2% R—Sq (adj) = 37.2%

Analysis of Variance

Source	DF	SS	MS	F	P
Regression	1	131.33	131.33	381.69	0.000
Error	643	221.24	0.34		
Total	644	352.58			

---

extreme caution. An examination of the residuals from this model suggests, with one or two exceptions, that they look reasonably normal, although with such a large sample (645), it is not surprising that a few aberrant values are present, for instance we would expect 5% (32) to fall outside of the 95% confidence limits.

At the other extreme of detecting a trend through time, we could fit log transformed arithmetic means to year. This is done in Table 8.4, a linear model gives a very good fit to the data with a R^2 value of 77%. A return to a mean below detection limit values is again suggested for the year 2003; 1961 is detected as a high residual value and distorts the residual diagnostic plots shown in Figure 8.12. This strongly suggests that something must be done to reduce the influence of the 1961 egg in this situation.

Table 8.4 Regression of log(*x* + 1) transformed arithmetic annual DDE means on year; 1961 is flagged as an outlier

---

Regression analysis

```
The regression equation is
logmndde = 6.01 - 0.0583 yr
```

Predictor	Coef	StDev	T	P
Constant	6.0117	0.4390	13.69	0.000
yr	-0.058293	0.005514	-10.57	0.000

S = 0.3311   R—Sq = 77.2%   R—Sq (adj) = 76.5%

Analysis of Variance

Source	DF	SS	MS	F	P
Regression	1	12.250	12.250	111.76	0.000
Error	33	3.617	0.110		
Total	34	15.867			

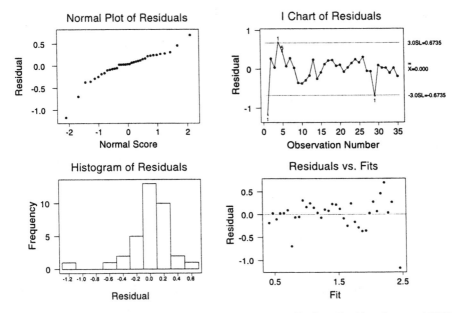

Figure 8.12 Residual diagnostic plots from the regression of $\log(x + 1)$ arithmetic annual DDE means on year.

The use of weights in regression can place more influence on certain points than on others. In this example we use the sample size in each year as the weight, so that the 1961 egg has less influence. The results of this weighted regression are shown in Table 8.5. R^2 has risen to 87% (not shown in the MINITAB output) and the equation would suggest a return of DDE mean value to non-detect levels in the year 2004. The egg from 1961 no longer has such an

Table 8.5 Weighted regression of $\log(x + 1)$ transformed arithmetic annual DDE means on year. Weights equal to sample size in each year

```
Regression analysis

Weighted analysis using weights in eggno

The regression equation is
logmndde = 6.15 - 0.0594 yr

Predictor      Coef        StDev        T          P
Constant       6.1505      0.3309       18.59      0.000
yr            -0.059411    0.004018    -14.79      0.000

Analysis of Variance

Source         DF          SS           MS         F          P
Regression     1           159.05       159.05     218.68     0.000
Error          33          24.00        0.73
Total          34          183.05
```

Table 8.6 Example regression of annual mean log(x + 1) transformed DDE on year. Compare with Table 8.3

Regression analysis

Weighted analysis using weights in eggno

The regression equation is
mnlogdde = 5.55 − 0.0540 yr

Predictor	Coef	StDev	T	P
Constant	5.5481	0.2969	18.69	0.000
yr	−0.053987	0.003605	−14.98	0.000

Analysis of Variance

Source	DF	SS	MS	F	P
Regression	1	131.33	131.33	224.31	0.000
Error	33	19.32	0.59		
Total	34	150.65			

---

influential residual. There is some suggestion of a tendency for more recent residuals to be positive (i.e. the model underestimates the data); if so this will affect our prediction of a return of DDE to below detection limit. It is thought that DDE residues will remain in the environment for some considerable time, so the year 2004 seems quite near for the mean residue level to drop below detection. It should, of course, be remembered that even if the mean falls below detection there will undoubtedly be eggs above detection limit for some further years. We do not have space here to discuss in detail when DDE will no longer be detected in any peregrine eggs.

The regressions that we have done here are slightly distorted by log transformation. Normally the weighted (by sample size) regression of annual mean on year will give exactly the same regression coefficients as for all the data (see Table 8.6). It can be seen that the degrees of freedom in the model are reduced; thus analysis using annual means giving us a conservative estimate of the number of independent data points.

Table 8.7 repeats the weighted regression on log mean HEOD. A significant decline over time is detected. Ignoring any reservations, for a moment, we can use the model to predict a return to mean levels equivalent to non-detection in the year 1997. We have some reservations about this model, however, because of the peak during the 1970s. An examination of residual diagnostic plots (not shown here) continues to suggest an elevated period of HEOD in the 1970s, which makes fitting a straight line questionable.

Our regression analysis of log mean PCB levels is summarized in Table 8.8 and a non-significant linear trend through time is suggested. However, the residual plots (Figure 8.13), indicate a non-random pattern of residuals and suggests a curved relationship. Figure 8.13 suggests a straight line relationship is dubious given the elevated levels in the 1980s.

Table 8.7 Weighted regression of log mean HEOD on year. $R^2 = 36\%$

Regression analysis

Weighted analysis using weights in eggno

The regression equation is
logmnheod = 1.86 − 0.0191 yr

Predictor	Coef	StDev	T	P
Constant	1.8637	0.3651	5.10	0.000
yr	−0.019127	0.004434	−4.31	0.000

Analysis of Variance

Source	DF	SS	MS	F	P
Regression	1	16.476	16.476	18.61	0.000
Error	33	29.215	0.885		
Total	34	45.691			

Unusual Observations

Obs	yr	logmnh	Fit	StDev Fit	Residual	St Resid
14	75.0	0.774	0.429	0.048	0.345	2.01R
15	76.0	1.250	0.410	0.045	0.840	3.75R

Table 8.8 Weighted regression of log mean PCB on year. $R^2 = 2\%$

Regression analysis

Weighted analysis using weights in eggno

The regression equation is
logmnpcb = 2.41 − 0.0089 yr

30 cases used 5 cases contain missing values or had zero weight

Predictor	Coef	StDev	T	P
Constant	2.406	1.107	2.17	0.038
yr	−0.00888	0.01331	−0.67	0.510

Analysis of Variance

Source	DF	SS	MS	F	P
Regression	1	2.746	2.746	0.44	0.510
Error	28	172.943	6.177		
Total	29	175.689			

Table 8.9 summarizes a quadratic response to the data. This is highly significant ($R^2 = 35\%$) and a major improvement on a linear model. Figure 8.14 shows the residuals plots and Figure 8.15 compares the linear and quadratic fits. There may be some suggestion that the pattern is asymmetric, but the quadratic curve forces the curves either side of the maximum to be symmetric. Some fractional polynomial curves are not restricted in this way. However, fitting a modified quadratic model of the form $y = a + bx + c\sqrt{x}$ (also shown in Table 8.9) only

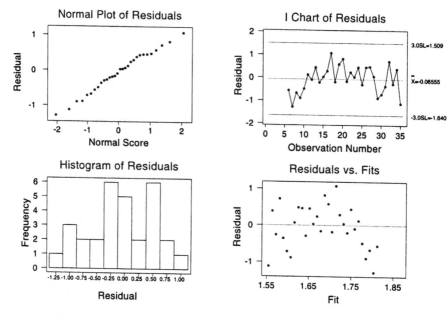

Figure 8.13   Residual diagnostic plots from the regression of $\log(x + 1)$ arithmetic annual PCB means on year. Note the curvature suggested in the residual charts.

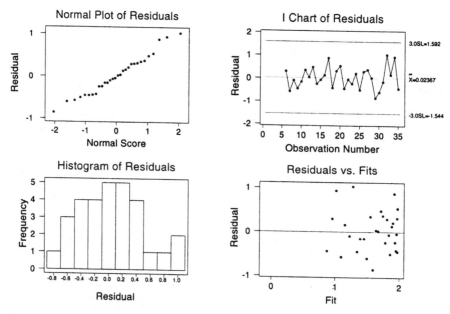

Figure 8.14   Residual diagnostic plots from the quadratic regression of $\log(x + 1)$ arithmetic annual PCB means on year.

Table 8.9 Weighted regression of log mean PCB on year. Quadratic and modified quadratic (replacing x^2 by $\sqrt{x}$ term). $R^2 = 35\%$ in both cases

---

Regression analysis

Weighted analysis using weights in eggno

The regression equation is
logmnpcb = − 31.2 + 0.817 yr − 0.00503 yrsq

30 cases used 5 cases contain missing values or had zero weight

Predictor	Coef	StDev	T	P
Constant	−31.204	9.166	−3.40	0.002
yr	0.8170	0.2244	3.64	0.001
yrsq	−0.005030	0.001365	−3.69	0.001

Analysis of Variance

Source	DF	SS	MS	F	P
Regression	2	60.629	30.315	7.11	0.003
Error	27	115.060	4.261		
Total	29	175.689			

Source	DF	Seq SS
yr	1	2.746
yrsq	1	57.883

Regression analysis

Weighted analysis using weights in eggno

The regression equation is
logmnpcb = − 132 − 1.65 yr + 29.7 sqrtyr

30 cases used 5 cases contain missing values or had zero weight

Predictor	Coef	StDev	T	P
Constant	−131.76	35.84	−3.68	0.001
yr	−1.6543	0.4395	−3.76	0.001
sqrtyr	29.748	7.943	3.75	0.001

Analysis of Variance

Source	DF	SS	MS	F	P
Regression	2	61.871	30.936	7.34	0.003
Error	27	113.818	4.215		
Total	29	175.689			

Source	DF	Seq SS
yr	1	2.746
sqrtyr	1	59.125

---

marginally improved the fit of the model and was virtually similar. The quadratic model suggests that a peak was reached in 1981 and that mean non-detect values could be reached early the following century. However, extrapolation from quadratic models is dangerous because the pattern is unlikely to hold far beyond the period considered.

Table 8.10 attempts to fit a trend line to the mean mercury data. This is unsuccessful, suggesting no recent trend through time.

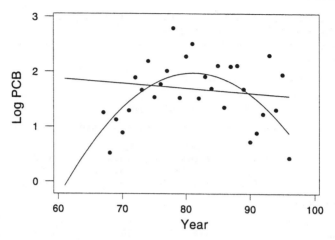

Figure 8.15 Linear and quadratic regression lines fitted to the log(x + 1) arithmetic annual PCB means.

Table 8.10 Weighted regression of log mean Hg on year. R^2 = 11%. No significant trend detected

---

Regression analysis

Weighted analysis using weights in eggno

The regression equation is
logmnhg = 1.82 − 0.0146 yr

15 cases used 20 cases contain missing values or had zero weight

Predictor	Coef	StDev	T	P	
Constant	1.825	1.051	1.74	0.106	
yr	−0.01456	0.01163	−1.25	0.233	

Analysis of Variance

Source	DF	SS	MS	F	P
Regression	1	0.6619	0.6619	1.57	0.233
Error	13	5.4887	0.4222		
Total	14	6.1506			

For the analysis of shell index, which is approximately normally distributed, we have no need to log transform the data. Table 8.11 and Figure 8.16 summarize the regression model and residuals diagnostic checking of the linear model. The latter give some suggestion of curvature in recent years. This would fit with our expectation of the data, i.e. that shell index will level out at pre-DDT levels. This combination of empirical evidence and conceptual evidence in model building is termed eclectic modelling (Gilchrist 1984) and is likely to produce much more acceptable models in the long term. However, in this example the

Table 8.11 Weighted linear and quadratic regressions of mean shell index on year. $R^2 = 67\%$ and 70% respectively. The quadratic is not a significant improvement on the fit of the model

---

Regression analysis

Weighted analysis using weights in eggno

The regression equation is
mnsi = 1.00 + 0.00881 yr

34 cases used 1 cases contain missing values or had zero weight

Predictor	Coef	StDev	T	P
Constant	1.00024	0.09031	11.08	0.000
yr	0.008809	0.001095	8.05	0.000

Analysis of Variance

Source	DF	SS	MS	F	P
Regression	1	2.4628	2.4628	64.76	0.000
Error	32	1.2170	0.0380		
Total	33	3.6798			

Regression analysis

Weighted analysis using weights in eggno

The regression equation is
mnsi = − 0.267 + 0.0405 yr − 0.000196 yrsq

34 cases used 1 cases contain missing values or had zero weight

Predictor	Coef	StDev	T	P
Constant	−0.2675	0.7751	−0.35	0.732
yr	0.04046	0.01926	2.10	0.044
yrsq	−0.0001957	0.0001189	−1.65	0.110

Analysis of Variance

Source	DF	SS	MS	F	P
Regression	2	2.5607	1.2803	35.46	0.000
Error	31	1.1192	0.0361		
Total	33	3.6798			

Source	DF	Seq SS
yr	1	2.4628
yrsq	1	0.0978

---

inclusion of a quadratic term gives only a minor improvement to the fit (Figure 8.17) suggesting the use of a quadratic term is not necessary. The linear model suggests that, if recent trends continue, the mean shell index should have returned to the pre-DDT value of 1.82 (Ratcliffe 1980) in 1993; a date that has already passed. The quadratic suggests a peak of 1.82 should be reached in the year 2003 and a modified quadratic (using square root x rather than x^2) suggest a peak of 1.83 should be reached in 2007. The best estimate probably lies somewhere between these alternatives. We anticipate that shell index will level

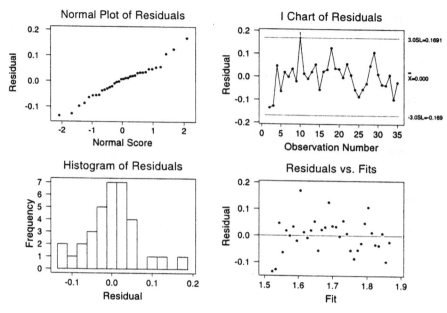

Figure 8.16 Residual diagnostic plots from the regression of arithmetic annual shell index means on year.

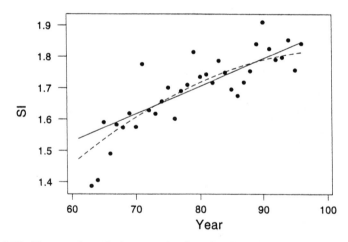

Figure 8.17 Linear and quadratic regression lines fitted to the arithmetic annual shell index means. Year displayed as years post-1900.

off, so neither linear nor quadratic models are adequate outside of the current range of the data and can only be seen, indeed as they should be seen, as mathematical approximations to reality.

An alternative approach to examine egg shell index would be to use logistic regression (see Chapter 5) on the proportion of eggs achieving an index value

Table 8.12 Logistic regression on the number of eggs each year achieving shell index $\geqslant 1.82$

Binary logistic regression

Link Function: Logit

Response Information

Variable	Value	Count
C118	Success	177
C119	Failure	341
	Total	518

34 cases were used
1 cases contained missing values

Logistic Regression Table

Predictor	Coef	StDev	Z	P	Odds Ratio	95% CI Lower	Upper
Constant	-7.060	1.124	-6.28	0.000			
C117	0.07725	0.01339	5.77	0.000	1.08	1.05	1.11

Log-Likelihood = -314.018
Test that all slopes are zero: G = 37.238, DF = 1, P-Value = 0.000

---

of 1.82 or greater. Figure 8.18 shows the proportions of eggs with shell indices $\geqslant 1.82$ for each year. Remember the varying sample sizes here when interpreting this figure. A logistic regression of the number of high index eggs ($\geqslant 1.82$) is given in Table 8.12 and the fitted curve added in Figure 8.18. The method predicts that 50% of sampled eggs should have indices $\geqslant 1.82$ in the

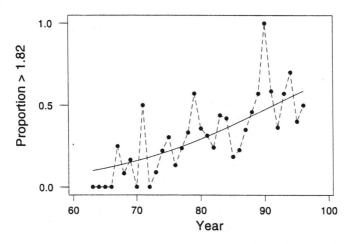

Figure 8.18 Proportion of eggs in each year (post-1900) achieving shell indices $\geqslant 1.82$, overlain with fitted line from logistic regression. Note that each year's proportion is based on a varying number of eggs.

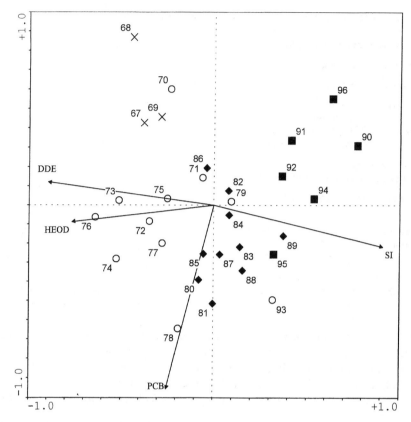

Figure 8.19 Principal component analysis biplot on log annual mean HEOD, DDE, PCB and annual mean shell index. This summarizes many of the trend features described earlier in this section (key: 1960s ×, 1970s ○, 1980s ◆, 1990s ■).

year 1991 (again, already past), and that 75% of eggs will achieve indices ⩾ 1.82 by the year 2006.

A summary of all the data in a Principal Components Analysis (PCA) biplot is given in Figure 8.19. This is based on log transformed annual means. Mercury has been excluded because of the reduced number of years. One of the features of PCA is that it cannot cope with missing values; hence 1961–66 are also excluded because there were no PCB measurements then (PCBs were not known as an environmental problem until the late 1960s). Figure 8.19 shows a clear temporal pattern from the 1960s through to the 1990s. This is in the direction of the shell index vector and away from DDE and HEOD vectors, confirming a trend towards higher shell index and lower DDE and HEOD values in the later years. The PCB vector is at right angles to the other three vectors and has associated with it a group of years including the 1980s, thus confirming the peak of PCB contamination in this period.

8.5 GEOGRAPHIC VARIATION

In an earlier study of peregrine eggs, covering the period until 1986, Newton *et al.* 1989, Figure 1, considered eight geographic areas from which eggs had been collected following Ratcliffe (1980). Mean values from these eight areas are summarized in Table 8.13 which reveals some interesting features. There is a predominance of eggs from northern England and Scotland reflecting the distribution of peregrines and their observers. Contaminant levels and shell indices are generally better (i.e. low and high respectively) in southwest England, Wales and the central and eastern Highlands. PCB and mercury levels are particularly high in the northern and western Highlands and DDE is high in southern Scotland. We do need to be cautious in interpreting these features as we have already shown that there are marked trends in these variables over time and we need to see whether there is a trend in time in the geographical collection of eggs.

Table 8.14 suggests that there has been a shift in collection over time. This table may be more easily examined by conversion to percentage contribution to each decade. This has been done in Table 8.15. It can be seen that south-west

Table 8.13 Area arithmetic means for the contaminants plus shell index

Area	*n*	HEOD	DDE	PCB	Hg	SI
1 South-west England	12	0.16	1.17	4.01	0.47	1.87
2 Wales	41	0.27	1.92	4.35	0.85	1.78
3 Northern England	123	0.37	3.27	3.14	0.73	1.77
4 Southern Scotland	191	0.69	4.68	4.64	0.56	1.69
5 Southern Highland fringe	95	0.39	3.51	8.96	1.14	1.68
6 Central and eastern Highlands	125	0.16	1.49	2.36	0.22	1.75
7 Northern and western Highlands	34	0.12	2.53	13.41	1.74	1.67
8 Ireland	24	0.33	3.20	6.81	1.03	1.65
Overall	645	0.40	3.21	5.12	0.69	1.72

Table 8.14 Numbers of eggs collected from each area by decade

Area	1960s	1970s	1980s	1990s	Overall
1 South-west England	0	0	6	6	12
2 Wales	0	6	9	26	41
3 Northern England	9	22	64	28	123
4 Southern Scotland	29	75	57	30	191
5 Southern Highland fringe	4	31	38	22	95
6 Central and eastern Highlands	19	20	63	23	125
7 Northern and western Highlands	2	14	14	4	34
8 Ireland	3	0	21	0	24
Total	66	168	272	139	645

Table 8.15 Area contributions (%) to decadal egg collections

Area	1960s	1970s	1980s	1990s	Overall
1 South-west England	0	0	2	4	2
2 Wales	0	4	3	19	6
3 Northern England	14	13	24	20	19
4 Southern Scotland	44	45	21	22	30
5 Southern Highland fringe	6	18	14	16	15
6 Central and eastern Highlands	29	12	23	17	19
7 Northern and western Highlands	3	8	5	3	5
8 Ireland	5	0	8	0	4
Total	100	100	100	100	100

Table 8.16 Least square means (year adjusted means) of contaminants and shell index

Area	HEOD	DDE	PCB	Hg	SI
1 South-west England	0.38	2.63	2.51	0.57	1.82
2 Wales	0.46	3.52	2.88	0.82	1.68
3 Northern England	0.42	2.87	1.97	0.73	1.72
4 Southern Scotland	0.44	3.08	2.32	0.64	1.69
5 Southern Highland fringe	0.39	3.02	4.13	1.14	1.64
6 Central and eastern Highlands	0.20	1.25	1.37	0.35	1.73
7 Northern and western Highlands	0.08	2.06	5.92	1.30	1.65
8 Ireland	0.43	2.71	2.74	0.87	1.62

England and Wales in particular are dominated by records from recent decades (in fact, reflecting a recovery from the much depressed numbers of peregrines in these areas); hence their means would be expected to be lower because of declines in HEOD and DDE over time and by improvements in the shell index. Levels in southern Scotland may be exaggerated because of a higher proportion of eggs from the 1960s.

The difficulty is in knowing how to make a valid comparison between these areas when their temporal pattern of collection was different. In Table 8.16 least squares means are presented from an analysis of variance including year and area terms, these can be considered as year adjusted means. Except for shell index these are back conversions from $\log(x + 1)$ transformed data. They can be compared to the means from Table 8.13. For HEOD and PCB, areas 6 and 7 (northern Scotland) appear lowest, although area 7 (north-west Highlands) has the highest mean PCB levels. Shell index values for south-west England still seem the highest.

To clarify whether any differences do exist, residuals from HEOD, DDE, PCB and SI were calculated from ANOVA removing year effects. These residuals were then entered into a PCA. The resulting scatter, with 499 data points (all eggs with values of all variables) is confusing, but there does seem to be separation between the areas (Figure 8.20). This can be clarified by just presenting the centroids for each of the areas (Figure 8.21). Figure 8.21 suggests that areas 6 and 7 are distinct from the remainder; area 6 because of relatively low contaminant to shell index values and area 7 because of low HEOD to shell index values.

Table 8.17 examines the slope of log annual mean DDE on year for each of the eight areas. There is some suggestion here that the slope (decline in DDE)

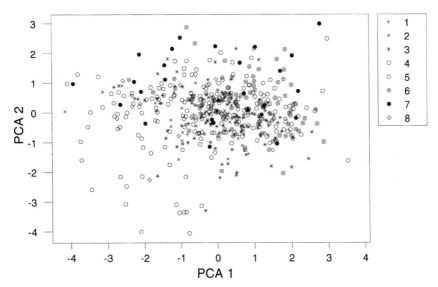

Figure 8.20 PCA plot based on residuals of HEOD, DDE, PCB and shell index after removing year effect. Each egg is denoted by a symbol denoting area of origin as in Table 8.13.

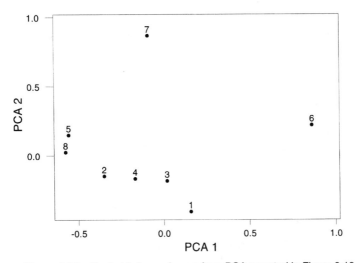

Figure 8.21 Centroids for each area from PCA reported in Figure 8.19.

has been greatest in areas 3 and 4 (northern England and southern Scotland). We can formally test whether these slopes are different using a modification of analysis of covariance. This is shown in Table 8.18 where the presence of different slopes would be indicated by a significant year (covariate)× area (factor) interaction. This clearly shows that these slopes are significantly different from one another, i.e. the test rejects the null hypothesis of the equality of slopes.

The results given in Table 8.18 suggest that the areas with high DDE contamination in the 1960s are reducing their burden more quickly than other

Table 8.17 Weighted regression of log(x + 1) transformed mean DDE on year for each area. Weights equal to sample size; n = number of years

Area	n	Slope	SE	P
1 South-west England	7	−0.063	0.014	0.006
2 Wales	12	−0.036	0.009	0.003
3 Northern England	30	−0.064	0.008	0.000
4 Southern Scotland	29	−0.076	0.005	0.000
5 Southern Highland fringe	22	−0.054	0.015	0.002
6 Central and eastern Highlands	28	−0.035	0.007	0.000
7 Northern and western Highlands	14	−0.041	0.019	0.056
8 Ireland	9	−0.020	0.018	0.300

Table 8.18 Analysis of covariance to test equality of slopes. Slopes are given as deviations from the mean slope, the deviation for area 8 (0.02890) is not shown

General linear model

```
Factor  Levels  Values
Area      8      1     2     3     4     5     6     7     8
```

Analysis of Variance for LogDDE

Source	DF	Seq SS	Adj SS	Adj MS	F	P
Year	1	142.121	18.972	18.972	38.11	0.000
Area	7	31.279	15.822	2.260	4.54	0.000
Area*Year	7	12.713	12.713	1.816	3.65	0.001
Error	135	67.211	67.211	0.498		
Total	150	253.323				

Term	Coef	StDev	T	P
Constant	5.1823	0.6899	7.51	0.000
Year	−0.048488	0.007855	−6.17	0.000

Year*Area				
1	−0.01456	0.04447	−0.33	0.744
2	0.01260	0.01831	0.69	0.493
3	−0.01556	0.01037	−1.50	0.136
4	−0.027605	0.009462	−2.92	0.004
5	−0.00521	0.01190	−0.44	0.662
6	0.013783	0.009992	1.38	0.170
7	0.00765	0.01673	0.46	0.648

Table 8.19 Mean contaminant and shell index values for each area for two periods (1961–80 and 1981–96); x indicates no data

Area	HEOD		DDE		PCB		Hg		SI		n_{max}	
	61–80	81–96	61–80	81–96	61–80	81–96	61–80	81–96	61–80	81–96	61–80	81–96
1	x	0.16	x	1.17	x	4.01	x	0.47	x	1.87	0	12
2	0.79	0.15	3.06	1.64	3.08	4.66	1.32	0.82	1.64	1.82	8	33
3	0.73	0.24	7.23	1.81	2.40	3.36	2.61	0.70	1.67	1.79	33	90
4	1.10	0.13	6.82	1.78	4.58	4.71	1.09	0.51	1.64	1.77	110	81
5	0.70	0.17	4.70	2.64	11.31	7.30	x	1.14	1.67	1.69	40	55
6	0.27	0.10	2.47	1.02	2.27	2.39	x	0.22	1.75	1.75	41	84
7	0.14	0.11	3.06	2.06	7.48	18.69	x	1.74	1.59	1.73	16	18
8	0.33	0.33	5.67	2.85	0.01	7.14	x	1.03	x	1.65	3	21
											251	394

Table 8.20 Mean contaminant and shell index values for non-coastal and coastal eggs

	HEOD	se	DDE	se	PCB	se	Hg	se	SI	se	n_{max}
Non-coastal	0.40	0.04	3.01	0.18	3.86	0.39	0.66	0.05	1.73	0.01	573
Coastal	0.44	0.06	4.84	0.48	14.87	2.64	1.32	0.41	1.64	0.02	72

areas. Table 8.19 is an attempt to examine this. Means are given for two periods, 1961–80 and 1981–96. With the exception of PCB, where a 1980s peak clouds the picture, there appears to be fewer differences between the areas in recent years than earlier in the monitoring program.

A small number of eggs were coastal in origin. Table 8.20 summarizes mean values over the whole period for non-coastal and coastal eggs. This presentation will be distorted by the temporal pattern of coastal egg collection. Few coastal eggs have been collected in the last decade, so means might be expected to be elevated. There are suggestions that PCB in particular is elevated.

Figure 8.22 displays annual mean PCB for non-coastal and coastal eggs. The peak of PCB shown previously for the early 1980s can be seen to be in large part attributed to coastal eggs. For non-coastal eggs there is a slight suggestion of a curved response through time, but neither this or a linear trend through time was detected. We would conclude that the apparent peak in PCBs is a result of coastal eggs being highly contaminated.

8.6 SHELL INDEX; INFLUENCING VARIABLES

Table 8.21 summarizes correlations between shell index and potential explanatory variables, log HEOD, log DDE, log PCB, log Hg and year (for trends). We have already seen that shell index has improved over time, while DDE values have declined. Thus it will be difficult to separate the effects of time on shell

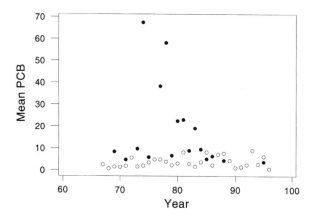

Figure 8.22   Annual mean PCB concentrations for non-coastal (○) and coastal eggs (●).

Table 8.21 Correlations between variables

```
Correlations (Pearson)

          SI        logheod   logdde    logpcb    loghg
logheod -0.167
logdde  -0.429      0.447
logpcb  -0.245      0.229     0.497
loghg   -0.109      0.211     0.418     0.453
YEAR     0.347     -0.391    -0.610    -0.094    -0.133
```

index from the effects of DDE. Using biological knowledge, rather than empirical evidence, we would expect DDE levels to influence shell index values. In regression, we can only assume association rather than causation. Here though, we bring together biological and empirical evidence and the synergy thus created is a good example of biologist and statistician working together. Of these correlations, log DDE is most linearly related to shell index, followed by year. Thus it would seem that the influence of log DDE concentrations on shell index is stronger than the year trend on its own. Normally we might investigate these patterns using stepwise regression (see Chapter 5), but here we have different sample sizes for the potential explanatory variables and we want to retain as many values (eggs) as possible in the analysis.

Table 8.22 summarizes the regression of shell index on log DDE. There are only 518 data points common to both shell index and DDE. The relationship is highly significant and the regression coefficient suggests that a unit change in log(DDE + 1) concentration would be responsible for a 0.121 change in shell index. We can be reasonably happy about the fit of these data; there is one odd residual, but otherwise residual diagnostic plots reveal no major problems.

Table 8.23 correlates the residuals from this model with other candidate

Table 8.22 Regression of shell index on log(x + 1) transformed DDE values

---

Regression analysis

The regression equation is
SI = 1.86 − 0.121 logdde

518 cases used 127 cases contain missing values

Predictor	Coef	StDev	T	P
Constant	1.85952	0.01488	125.00	0.000
logdde	−0.12133	0.01125	−10.79	0.000

S = 0.1801   R—Sq = 18.4% R—Sq (adj) = 18.2%

Analysis of Variance

Source	DF	SS	MS	F	P
Regression	1	3.7728	3.7728	116.34	0.000
Error	516	16.7340	0.0324		
Total	517	20.5069			

Table 8.23 Correlations of residuals from log DDE model with other candidate variables and regression of shell index on log(x + 1) DDE and year

---

Correlations (Pearson)

	RESI15
logheod	0.002
logdde	−0.000
logpcb	−0.066
loghg	0.052
YEAR	0.110

Regression analysis

The regression equation is
SI = 1.52 − 0.0970 logdde + 0.00379 YEAR

518 cases used 127 cases contain missing values

Predictor	Coef	StDev	T	P
Constant	1.5213	0.1106	13.76	0.000
logdde	−0.09703	0.01365	−7.11	0.000
YEAR	0.003787	0.001227	3.09	0.002

S = 0.1786   R—Sq = 19.9%   R—Sq (adj) = 19.6%

Analysis of Variance

Source	DF	SS	MS	F	P
Regression	2	4.0767	2.0384	63.89	0.000
Error	515	16.4301	0.0319		
Total	517	20.5069			

Source	DF	Seq SS
logdde	1	3.7728
YEAR	1	0.3039

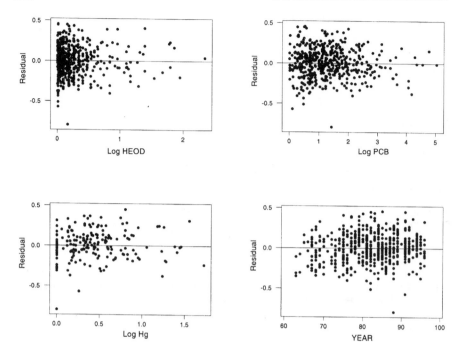

Figure 8.23 Plots of residuals from the regression of shell index on $\log(x + 1)$DDE against other candidate variables: $\log(x + 1)$HEOD, $\log(x + 1)$PCB, $\log(x + 1)$Hg and year.

variables. The correlation with year is most pronounced. In case there are any relationships that are not linear we have plotted residuals against potential additional variables in Figure 8.23. Our eye is drawn to the lone low residual, but there are no definite relationships suggested here. Table 8.23 then goes further by adding year into our regression model. We have managed to improve the fit of the model from an R^2 of 18.4% to one of 19.9%. This is a modest improvement but a significant one in view of the large dataset.

The regression in Table 8.23 might be interpreted in the following way. The regression coefficient for year is positive, that for log (DDE +1) is negative. These are partial regression coefficients, i.e. a coefficient calculated while keeping the other variable fixed. Thus we would expect, after removing year trend, that a unit change in log (DDE + 1) would be associated with a 0.097 change in shell index. This is a lower figure than before and reflects the removal of the influence of year. At a fixed level of log DDE the coefficient for year suggests that a 0.0038 improvement per annum in shell index has been taking place.

Table 8.24 now correlates the residuals from this model with other candidate explanatory variables. The relationship with log PCB is strongest and the result of including this in the regression model is summarized at the bottom of table 8.24. This has now reduced the number of eggs included in the analysis to 500,

Table 8.24 Correlations of residuals from log DDE + year model with other candidate variables and regression of shell index on log(x + 1) DDE, year and log (x + 1) PCB

```
Correlations (Pearson)

              RESI17
logheod       0.023
logdde       -0.000
logpcb       -0.100
loghg         0.030
YEAR         -0.000


Regression analysis

The regression equation is
SI = 1.51 - 0.0801 logdde + 0.00407 YEAR - 0.0259 logpcb

500 cases used 145 cases contain missing values

Predictor    Coef          StDev        T         P
Constant     1.5135        0.1202       12.59     0.000
logdde      -0.08012       0.01636      -4.90     0.000
YEAR         0.004072      0.001355      3.01     0.003
logpcb      -0.02591       0.01049      -2.47     0.014

S = 0.1773    R-Sq = 18.8%    R-Sq (adj) = 18.3%

Analysis of Variance

Source         DF    SS        MS       F       P
Regression     3     3.6145    1.2048   38.33   0.000
Error          496   15.5907   0.0314
Total          499   19.2052

Source         DF    Seq SS
logdde         1     3.2354
YEAR           1     0.1873
logpcb         1     0.1918
```

so it seems as if the R^2 has gone down. Had we started with these 500 eggs, R^2 for log DDE alone would have been 16.8%, rising to 17.8% on addition of year and to 18.8% as shown when log PCB was finally added. There is a suggestion that elevated PCB levels are detrimental to shell index. Neither log HEOD or log Hg improved the fit of the model shown.

8.7 SUMMARY

We have looked at a large monitoring data set. It has some typical and some atypical features associated with ecotoxicological data. The typical features include highly variable data, non-detection levels, skewed distributions with a few extreme points and missing data. There are also changes through time to geographical coverage. The sampling programme is far from ideal because of other constraints to a well-designed sampling programme. More unusual are the duration and size of the data set; the data shown here are the result of 36 years of monitoring. The number of variables is also small; a scheme starting today would probably contain a whole suite of determinations that were neither practical or cheap to obtain in the 1960s.

Trends through time for DDE, HEOD and shell index were easily determined from these data. This all looks promising for peregrine populations. Mean levels of shell index are now close to pre-DDT levels and we can predict continuing improvements in years to come. The pattern for mercury is less clear because of the shorter time period involved. Total PCB levels peaked in the early 1980s. Subsequent examination of the data suggest that this feature was heavily influenced by coastal eggs where peregrines were taking seabirds whose own PCB loads were derived largely from the marine environment. The lack of coastal peregrine eggs in recent years may reflect reduced populations in these locations as a result of elevated PCB levels, but for this chapter this must remain as conjecture. We have shown that a principal components biplot can summarize temporal changes.

Different areas of Britain are characterized by different contamination patterns. Areas recently recolonised appear to have lower levels, northern Scotland has a higher pattern derived in part from a larger number of coastal records. All areas have experienced a decline in DDE levels over time, and there is evidence that areas initially most contaminated have recovered more quickly. There is much less difference between regions now than there was at the start of the monitoring program.

Regression analysis of shell index suggests that the index is heavily influenced by log DDE levels and that there is also a residual year effect and a possible adverse effect of PCBs. No effect of HEOD was detected, but it should be noted that HEOD and DDE are positively correlated. The results are consistent with and extend the earlier analysis of Newton *et al.* (1986). Declines in contaminant levels have coincided with the recovery of peregrine populations, providing evidence that organochlorines were the causal factors in decline.

This case study demonstrates that interpretation of data can be very difficult, particularly where data are unbalanced and gappy. The role of teamwork between statistician and biologist is essential to maximize the value of data sets such as this.

REFERENCES

Cooke AS, Bell AA and Haas MB (1982) *Predatory Birds, Pesticides and Pollution*, Institute of Terrestrial Ecology, Huntingdon.

Crick HQP and Ratcliffe DA (1995) The peregrine, *Falco peregrinus* breeding population of the United Kingdom in 1991. *Bird Study*, **42**, 1–19.

Gilchrist W (1984) *Statistical Modelling*, Wiley, Chichester.

Newton I (1986) *The Sparrowhawk*, Poyser, Carlton.

Newton I and Galbraith EA (1991) Organochlorines and mercury in the eggs of golden eagles, *Aquila chrysaetos* from Scotland. *Ibis*, **133**, 115–120.

Newton I and Haas MB (1984) The return of the sparrowhawk. *British Birds*, **77**, 47–70.

Newton I, Bogan JA and Rothery P (1986) Trends and effects of organochlorine compounds in sparrowhawk eggs. *Journal of Applied Ecology*, **23**, 461–478.

Newton, I, Bogan JA and Haas MB (1989) Organochlorines and mercury in the eggs of British peregrines *Falco peregrinus*. *Ibis*, **131**, 355–376.

Newton I, Wyllie I and Asher A (1993) Long-term trends in organochlorine and mercury residues in some predatory birds in Britain. *Environmental Pollution*, **79**, 143–151.

Newton I, Dale L and Little B (in press) Trends in organochlorine and mercurial compounds in the eggs of British merlins *Falco columbarius*. *Bird Study*.

Ratcliffe AD (1980) *The Peregrine Falcon*, Poyser, Carlton.

Ratcliffe AD (1993) *The Peregrine Falcon*, Poyser, London.

Sheail J (1986) *Pesticides and Nature Conservation. The British Experience 1950–1975*. Clarendon Press, Oxford.

Sparks TH and Rothery P (1996) Resampling methods for ecotoxicological data. *Ecotoxicology*, **5**, 197–207.

9

Statistical Techniques for the Ecological Risk Assessment of Chemicals in Freshwaters

MARK CRANE[1], ALBANIA GROSSO[2] AND COLIN JANSSEN[3]
[1]*School of Biological Sciences, Royal Holloway, University of London, UK.*
[2]*WS Atkins Environment, Epsom, UK.*
[3]*Laboratory for Biological Research in Aquatic Pollution, University of Ghent, Belgium*

9.1 INTRODUCTION

The applied science of ecotoxicology has no meaning outside the context of chemical risk assessment. Ecotoxicologists, whether they work in the industrial, regulatory or academic sectors, do what they do because humans feel a need to know how, when and where chemicals affect nature. Recently there has been considerable activity in the field of human and ecological risk assessment, with detailed documents on best practice emerging from both North America (National Research Council 1993) and Europe (European Commission 1996). Risk assessment has been defined as the process of assigning magnitudes and probabilities to the adverse effects of human activities and natural catastrophes (Suter 1995). The ecological risk assessment of chemicals therefore involves the interpretation of data on the severity and likelihood of damage to organism assemblages from the release of chemicals by human activity. The word 'likelihood' strongly suggests that statistical tools should be used during this type of assessment. Suter (1995) describes the main elements of an ecological risk assessment (Figure 9.1).

In this chapter we will perform a risk assessment for a hypothetical substance likely to be released into the environment. We will use the assessment as a vehicle to illustrate the variety of statistical techniques that can be used within freshwater ecotoxicology, from simple descriptive summaries to more complex hypothesis tests and models. To maintain generality, we will simply refer to the substance as 'the substance', although knowledge of the physical and chemical

Statistics in Ecotoxicology. Edited by T. Sparks. © 2000 John Wiley & Sons Ltd

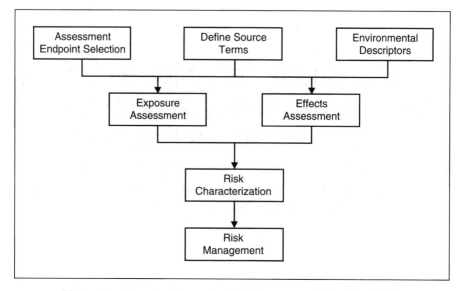

Figure 9.1   Elements of an ecological risk assessment (after Suter1995).

properties of a potentially hazardous substance is vital in most risk assessments. Our substance could be a pesticide, an effluent from an industrial plant, sewage treatment works or landfill, or a chemical used widely by industry.

The example will be a 'predictive' rather than a 'retrospective' risk assessment (Maltby and Calow 1989). In other words, the assessment is for a new substance and we will try to make some predictions about its likely environmental effects when released to the environment. Retrospective assessments (or 'impact assessments') will not be considered because of constraints on the length of this chapter. However, they are a vital and often neglected component in risk assessment (Fox 1991; Gilbertson *et al.* 1991; Power and McCarty 1997).

9.2   ASSESSMENT ENDPOINTS

Assessment endpoints should be chosen in phase one of a risk assessment. These are the entities that the risk assessor wishes to protect from harm. They will usually have both societal and biological relevance, so that the public and the scientific community will be prepared to support the risk assessment objectives. They should also have an unambiguous and concrete definition, so that everyone knows what it is that the risk assessor is trying to protect. They should be susceptible to the hazardous agent. Finally, they should either be accessible to direct measurement, or linked in some predictable way with an entity that can be measured (Warren-Hicks *et al.* 1989).

9.2.1 OPERATIONAL CONCEPTS

No formal statistical techniques are required when deciding upon assessment endpoints in an ecological risk assessment. However, it is vital for subsequent statistical analysis and for the success of the assessment that unambiguous endpoints are selected (Chapman *et al.* 1996). The term 'operationalization' has been used to describe the practical and precise specification of study objectives. Unfortunately, some concepts are impossible to operationalize because they are vague and amorphous. 'Ecosystem health' is an example of an unoperational concept that is frequently encountered in ecotoxicology (Crane and Newman 1996). A quantitative risk assessor would be ill advised to select ecosystem health as an assessment endpoint because, (1) there is little agreement on what the term means, and therefore (2) it is hard to arrive at a coherent series of measurements that might show whether or not a particular ecosystem is healthy (Calow 1992). Other examples of unoperational concepts are 'ecological community', 'biotic integrity' and 'ecosystem damage'. Instead of these, a risk assessor should identify meaningful assessment endpoints that can be specified accurately and precisely and are amenable to statistical quantification.

Reckhow (1994) recommends the construction of an 'objectives hierarchy' as an effective method for identifying the most important objectives within a decision analytic framework. Box 9.1 shows a simplified objectives hierarchy for the chemical risk assessment that we are performing. The overall aim at the top of the hierarchy is to manage the release of the substance so that no unacceptable damage to the environment occurs. Beneath this aim are several issue-specific objectives. Below this are the assessment endpoints, and below these the measurement endpoints. Measurement endpoints, also known as 'attributes' to decision analysts, are the things that can be measured and which reveal something useful about the assessment endpoints. Sometimes a measurement endpoint is easy to identify from an assessment endpoint. For example, if the assessment endpoint is the maintenance of viable trout populations then direct measurements of chemical effects on trout are one obvious measurement endpoint. In contrast, the measurements that should be made to ensure no effects on the biological cycling of nutrients are less clear and will therefore probably involve several surrogate measures.

Later in this chapter we take only one measurement endpoint, the prediction of hazard and risk to fish through laboratory toxicity testing, and use this as a specific example in several statistical exercises. All assessment endpoints and their many associated measurement endpoints should be considered in a real risk assessment, making this a complex and lengthy procedure.

9.3 SOURCE TERMS

Source terms should also be estimated in the first phase of a risk assessment. These are the spatial and temporal release patterns of chemicals from their

---

Box 9.1 Simplified objectives hierarchy for a freshwater ecological risk assessment

Manage the biological effects of substance release
  Maintain successful recreational fishery
    Maintain game fish populations
    Maintain coarse fish populations
      Bioassays with selected fish species (predictive)
      Fish population surveys (retrospective)
Meet water quality standards
    Meet List I standards
    Meet List II standards
    Meet water quality objectives
      Hydrological modelling of likely chemical fate and behaviour (predictive)
      Chemical surveys and analysis of specific determinands (retrospective)

  Minimize biological effects
    Maintain viable populations of naturally occurring aquatic plants and animals
    Maintain normal cycling of nutrients
    Maintain normal trophic status
    No encouragement of non-native invasive species
      Bioassays with representative species (predictive)
      Nitrification and biodegradability studies (predictive)
      Structural and functional biological surveys (retrospective)

  Maintain aesthetic conditions
    No nuisance smells
    No algal blooms
    No visible fish kills
    No increase in swarming insects
      Odour tests (predictive)
      Algal and fish bioassays (predictive)
      Pilot field studies (predictive)
      Field surveys (retrospective)

  Minimize financial costs
      Costs to discharger of increased research, restrictions on release, or improvement of treatment facilities (predictive and retrospective)
      Costs to regulator of increased research or monitoring (predictive and retrospective)

(Key to hierarchy: ***Overall aim***, **issue-specific objectives**, assessment endpoints, *measurement endpoints*)

---

sources. At the simplest level this could be from a continuous point source releasing effluent of known composition, such as may occur from a sewage treatment works. On the other hand, the source term could be both diffuse and intermittent, such as may occur when agrochemicals are applied to land near freshwaters. The manufacturer or discharger of the substance that we are assessing should provide information on the amount that will be released into the environment over time. Often this information will be limited and a modelling exercise will have to be undertaken to provide some quantitative information on contaminant distributions for the risk assessor to work with. Discussion of appropriate modelling techniques is beyond the scope of this chapter, but the interested reader can find more information on the subject in Mackay (1991) or Cowan *et al.* (1995).

How can numerical information supplied on contaminant sources be summarized effectively? Let us imagine that the manufacturer of our substance has already performed a pilot or desk study in which the release of the substance during typical use was measured or modelled through time. We will begin by summarizing the data using statistical descriptions that are familiar to all readers, starting with simple parametric summaries and tests based upon the normal distribution discussed in earlier chapters. We will then move on to simple non-parametric statistics in which the distribution of data is either unknown or clearly not normal. The data, summaries and analyses used in this section can all be found in Box 9.2.

Box 9.2 Description of source terms

Concentration of substance released into the environment recorded in a pilot study at 12.00 hours each day over a three week period ($n=21$):

14, 5, 10, 15, 6, 18, 9, 21, 17, 13, 12, 12, 4, 16, 19, 22, 23, 19, 10, 14, 16 mg l^{-1}.

Mean, $\bar{x}$,	$= 14.05$ mg l^{-1}
Median	$= 14$ mg l^{-1}
Sample variance, s^2	$= 29.45$
Standard deviation, s	$= 5.43$
CV	$= 39\%$

Standard error of the mean $= s/\sqrt{n} = 1.18$
95% confidence interval for the mean: calculated as $\bar{x} \pm t_v\,s/\sqrt{n}$
$t_{20}=2.086$, hence lower 95% confidence limit $=14.05-2.086 \times 1.18$ and upper 95% confidence limit $=14.05+2.086 \times 1.18$, i.e. 95% confidence interval (11.58, 16.52).

9.3.1 DESCRIPTIVE STATISTICS

Ways of summarizing data with different distributions were described in earlier chapters. A 'measure of dispersion' is usually calculated, such as the sample variance (s^2) or standard deviation (SD). The coefficient of variation (CV) is another useful descriptive statistic because it provides a measure of the sample variation relative to the mean. We might want this information if we wish to compare the variability of two data sets with different means. The CV is simply the standard deviation divided by the mean.

Another very important statistic is the standard error of the mean (SE or SEM), which is a measurement of error associated with our estimate of the true population mean. The size of a confidence limit associated with any statistical estimate will depend upon:

- the sample size: the bigger the sample, the smaller the error term
- the variability of the data: the larger the SD, the poorer the precision of the estimate
- the level of confidence we wish to have that the estimate does in fact lie within the calculated interval

It is conventional to set our level of confidence at 95%, so that 19 times out of 20 the population mean, for example, will be within the specified confidence interval. We call this the 95% confidence interval. The upper and lower bounds of this interval are called the 95% confidence limits.

We now have some summaries of our source terms. These are often used by a risk assessor without any further elaboration. For example, the upper 95% confidence limit may be used as a conservative estimate of the amount of contaminant that is likely to be released into the environment. By using this value, rather than a mean or a median, pressure is exerted upon the dischargers of chemicals to carry out further surveys, experiments and modelling exercises that generate more *precise* results, to help deter over-conservative regulation. However, we will show later how these data on contaminant concentrations can be more fully used to estimate the probability of harm.

9.3.2 COMPARISON OF SEVERAL DATA WITH A NOMINAL VALUE

Risk assessors may wish to compare a set of data with a standard value. In our example the assessor may wish to make comparison with a 10 mg l^{-1} standard. A one-sample t-test is a simple parametric method for comparing the pilot study data with this standard. Box 9.3 shows how this should be done. We encountered the t-statistic above when we looked at the calculation of a 95% confidence interval for sample sizes < 30. The underlying assumptions of the t-test are that the samples are drawn from populations of normally distributed, continuous variables.

Tables of critical values provide thresholds for either one-tailed or two-tailed t-tests. In our example, if we have a clear prior reason for believing that the recently collected data should have a mean that is significantly greater than the standard, we could justifiably use a one-tailed test. However, in many cases we do not have a good reason for believing in which direction ('more' or 'less') the

---

Box 9.3   One sample t-test and non-parametric test

One sample t-test $t_{20} = 3.42, p < 0.001$
so there is evidence that the new data come from a population with a mean greater than the standard.

Power of test

$$t_{\beta (2), \upsilon} = \delta / \sqrt{(s^2/n)} - t_{\alpha, \upsilon}$$

where

$t_{\beta (2), \upsilon}$ = power of t-test, δ = difference between means
$t_{\beta (2), \upsilon} = 4.05 / 1.18 - 1.725 = 1.707$

From tables of critical t-values, for $\upsilon = 20$, $\beta \equiv 0.05$. Therefore power $(1 - \beta) \equiv 95\%$.

Non-parametric test
The sign test eliminates ties (i.e. days when concentrations are equal to 10), and calculates the number of values below (5) and above (14) the standard of 10. Probabilities associated with the binomial distribution with parameter, p, equal to 0.5 suggest that a one-tailed significance test of $p = 0.001$ is achieved. A confidence interval for the median can also be derived as (11.35, 17.33) slightly wider than that based on the t-distribution.

We conclude, from both parametric and non-parametric tests that the sample exceeds a standard of 10mg l^{-1}.

---

change might be. In that case we have to take the more conservative two-tailed approach.

If the sample is small or clearly not normal, a non-parametric equivalent should be performed. Non-parametric alternatives to the one-sample t-test include the sign test and Wilcoxon. These non-parametric tests are strictly tests of medians, not means.

9.3.3   MINIMIZING STATISTICAL ERRORS

In statistical testing we need to decide what level of risk we are willing to accept of falsely rejecting the null hypothesis of 'no difference'. This level of risk is called the significance level and is normally rather small, by convention <0.05, or 5%. Concluding that data are different when they are not, is called a Type I error. Concluding that they are the same when they are not is called a Type II error. Clearly a risk assessor does not want to commit either type of error. The way to minimize Type I errors is to set a small significance level, an idea that most readers will be familiar with. What is less familiar to many is the minimization of Type II errors by maximizing the power of the statistical test.

Power is simply the ability of a statistical test to detect a difference between data when there actually is one. It is adversely affected when data are variable or samples are small, as is often the case in environmental studies. The only way to maximize both the significance and power of any statistical test is to increase the sample size. It is vital that a risk assessor knows the power of any statistical test performed, especially when that test suggests that there is *no* significant difference between sets of data. Box 9.3 provides a power calculation for the one-sample *t*-test performed on the data. It shows that in this example the *t*-test had a reasonable level of power to detect any difference between the standard and the pilot study data. Where data are normal, non-parametric tests are usually <95% as powerful as parametric alternatives (Zar 1984).

9.3.4 GEOSTATISTICS

The substance that we are assessing may have been used in another country, or data from pilot studies may be available that help us to predict the way in which it will disperse from a source point. If a risk assessor is expected to map the location of contaminants as well as measure their concentrations then geostatistical techniques are likely to be the most cost-effective tools. They are regularly used in the assessment and remediation of contaminated land (DoE 1994; Flatman and Yfantis 1996), but have not yet been used extensively in freshwater assessments.

Geostatistics differ from the 'classical' or 'frequentist' techniques discussed so far because they are designed to deal with environmental samples that are not independent but are, instead, spatially correlated (Cressie 1990). This is often the case, particularly when samples are solids such as aquatic sediments: samples taken close together will almost certainly be more similar than those that are taken further apart. This spatial correlation can be used to improve the power of statistical analysis by using a technique called kriging to analyse samples taken along a regular grid pattern. A 'semivariogram' is first constructed. This is a graphical tool for determining the distance and direction of sample correlations, which then provides information on the number of samples required for accurate kriging. Kriging is an interpolation technique that is used to estimate unknown points from surrounding data and then produce contour maps from the estimates.

Much has been written on geostatistics (e.g. Isaaks and Srivastava 1990; Journel 1986), so in this section we simply show in Box 9.4, Figs 9.2 and 9.3, how data on the distribution of our substance in sediments near a release point can be geostatistically analysed using readily available software such as Surfer 6.0 (Golden Software Inc. 1997). We should also point out that geostatistics can be used on data other than chemical concentrations. For example, toxic hazards and risks can be kriged (Ginevan and Splitstone 1997). If a picture is worth a thousand words in most walks of life, it must be worth considerably more in risk assessment where largely innumerate managers and members of the general public must be satisfactorily informed. Maps, graphs and diagrams are excellent

Box 9.4   Use of kriging to generate a map of contamination

83	89	90	97	101	103	100	98	95
107	109	113	117	119	120	120	119	115
132	137	140	139	140	145	140	143	141
137	141	145	150	150	150	150	148	148
141	143	147	150	160	166	158	155	154
145	148	152	157	164	171	167	161	160
151	154	158	162	170	175	172	169	164
155	159	163	169	175	180	177	170	165
157	164	170	174	180	184	181	175	170
163	167	175	180	185	190	187	181	177
165	169	180	189	192	195	190	183	180
169	173	181	190	200	200	200	189	180

Discharge Point

Figure 9.2   Data on spatial distribution of substance (mg/kg^{-1}) in aquatic sediments near discharge point.

Continued overleaf

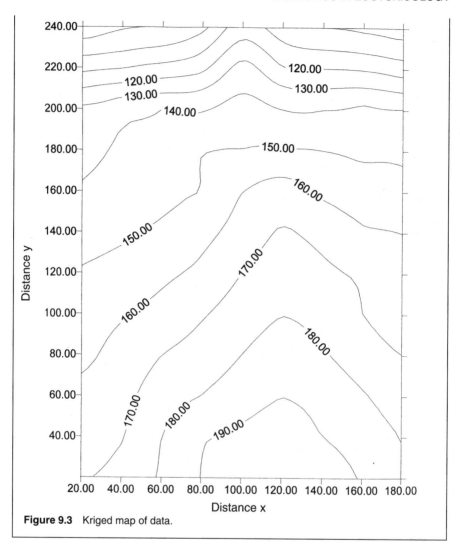

Figure 9.3 Kriged map of data.

tools for communicating hazard and risk, and should be used whenever possible.

9.4   ENVIRONMENTAL DESCRIPTION

The third component in the first phase of a risk assessment involves the definition of spatial and temporal boundaries around the environment that one wishes to protect. For new chemicals this phase is often rather abstract: the risk assessor usually wants to protect all environments that might possibly be exposed to the chemical. For site-specific or retrospective risk assessments this phase may be more concrete. In either case, the source term definition will depend partly upon the assessment endpoints chosen in phase 1. For example, if protection of trout populations is important then some understanding of trout movement, reproductive behaviour and life cycle would be useful in determining assessment boundaries.

The environmental description in our predictive risk assessment is necessarily generic because we cannot be certain where the substance will end up in the environment. However, information on its physicochemical properties, such as its solubility or partition coefficient, may suggest the substance will generally end up bound to organic material in aquatic sediments. It would then be useful to know whether sediment organic content differs substantially between different freshwater systems that may be exposed to the substance. This is because important differences in the characteristics of receiving environments may require us to describe them separately and perform different risk assessments for each.

In this section we will show how two samples, each comprising several data points, can be compared. We will also show how skewed data may be transformed to allow parametric analysis and how they can be displayed graphically.

9.4.1   COMPARISON OF TWO SAMPLES

Box 9.5 and Fig. 9.4 show some data on the organic carbon content of 20 randomly selected ponds and 20 river depositional zones in the region where the substance may be released. A pooled variance, two-sample t-test can be used to compare the means of two samples if (1) they both come from a normal distribution and (2) they both have equal variances. If the variances of the two populations are unequal, then it is incorrect to pool the variance. Instead an alternative t-test should be performed, but the calculations remain similar. An F-test is one way of testing whether variances are equal. The non-parametric alternative to the two-sample t-test is the Mann–Whitney test.

9.4.2 TRANSFORMATIONS

Environmental data are often positively skewed, with a bunching of low values and a spread to very high values (Davies 1989). These data may be made to approximate a normal distribution by transforming them. There are many different ways of transforming data, including arcsin and square root transformations for percentage/proportion data. The data in our example are percentages, and the pond data appear log-normally distributed. In this example

Box 9.5 Comparison of organic carbon content in the sediments of ponds and rivers

Data set (% organic carbon contest)

River	3	6.3	5.9	2.3	4.5	4	5.9	6.3	6.6	4.5	4.8	6.6	6.3
Pond	17.9	6.3	15.9	12.6	4.5	40	5.9	6.3	66.2	4.5	25.8	33.7	6.3

River	5.4	6.1	3	7	4.2	5.8	5.7
Pond	53.8	6.1	30.1	7	14.2	5.8	5.7

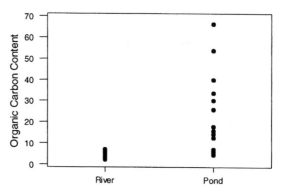

Figure 9.4 Data scatter.

	River	Pond
Mean	5.21	18.43
Mean Log_{10} transformed	0.7	1.1
Median	5.75	9.8
Variance (s^2)	1.84	319.5
Var Log_{10} transformed	0.018	0.148
SD	1.36	17.87
SE	0.30	3.997
CV %	26	97

Using *F*-tests to check for equality of variance we obtain (largest variance as numerator) 319.5/1.84 = 173.64 and 0.148/0.018 = 8.22 for untransformed and transformed data respectively. The transformation has not been successful in equalizing variance. Hence here we used a *t*-test, without pooling the variance.

Two-sample t-test $t = -4.35, p < 0.001$, two tailed test, log transformed data

Mann–Whitney test $p < 0.001$

Both tests conclude that a difference exists between the two water body types. From the statistical analyses used in our environmental description we now know that separate risk assessments may be required for ponds and rivers because they form two distinctive groups.

we have used a $\log_{10}$ transformation. The value of transforming data is that more powerful parametric analyses can be used on data that in their original form violate the assumptions of parametric statistics.

9.4.3 GRAPHICAL PRESENTATION

It is often useful to group environmental data to produce a frequency distribution or to present the data by other graphical means, such as boxplots. When the data are plotted as a bar chart (discrete or nominal data) or histogram (continuous data) it is possible to determine the 'shape' of the data (e.g. normal or not) and the presence of outliers that may require further investigation.

9.5 EXPOSURE ASSESSMENT

The second phase of a risk assessment is among the most difficult for the risk assessor and involves the conversion of the source term into an estimate of the extent to which receptor organisms are actually exposed. This is often referred to as an 'exposure assessment.' In our example we already know that organic carbon in sediment is likely to affect the partitioning of the substance we are assessing. Therefore it is also likely that the presence of dissolved organic carbon (DOC) in the water will influence the bioavailability of the substance to organisms such as fish: the greater the concentration of DOC the lower the exposure. Laboratory tests can be used to establish a relationship between contaminant uptake by aquatic organisms and the amount of DOC in surrounding water. One way of visualizing this type of relationship is through a simple scatterplot. Box 9.6 shows a scatterplot of DOC versus tissue concentrations of the substance in fish, Fig. 9.5. Regression is the parametric technique most often used to analyse this type of data. Spearman's rank correlation coefficient (r_s) can be used to look for association in non-normal data.

9.5.1 LINEAR REGRESSION

In simple linear regression we expect to find a linear relationship between two variables, one of which we believe is dependent on an independent variable. The independent variable contributes the x-values along the horizontal axis of a scatterplot. The dependent variable consists of the data that we hope have a functional dependence on x and contributes the y-values along the vertical axis of a scatterplot. The y-values should be normally distributed for every value of x. The variance of y at each value of x should also be the same. In our example

Box 9.6 Regression and Spearman's rank correlation to find associations between dissolved organic carbon concentrations and concentrations of the substance in fish

Dataset

DOC (mg l⁻¹)	5	9	19	24	26	42	58	71	93	110	131	155
Tissue concentration (mg g⁻¹)	48	46	41	40.6	40	38	35.5	33	30.1	28.8	26	23.9

DOC (mg l⁻¹)	181	200	214	233	260	277	285	290
Tissue concentration (mg g⁻¹)	21.2	20.3	19	16.1	15.9	15	14.8	14.6

Regression analysis

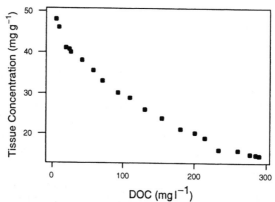

Figure 9.5 Scatterplot of tissue concentration against DOC.

The regression equation is
Tissue Concentration (mg g⁻¹) = 42.9 - 0.108 DOC (mg l⁻¹)

Predictor	Coef	StDev	T	P
Constant	42.8650	0.9327	45.96	0.000
DOC	-0.107901	0.005593	-19.29	0.000

S = 2.478   R-Sq = 95.4%   R-Sq(adj) = 95.1%

Analysis of Variance

Source	DF	SS	MS	F	P
Regression	1	2286.6	2286.6	372.25	0.000
Error	18	110.6	6.1		
Total	19	2397.2			

DOC is the independent variable and tissue concentration the dependent variable. Polynomial regression and multiple regression techniques are also available if the relationship is non-linear or involves more than two variables (see Chapter 5).

A regression line shows the linear regression of y on x: it explains how y changes with values of x and is summarized by the well-known equation $y = a + bx$. In this equation b is the slope of the line (or regression coefficient) and can be positive or negative; a is where the line intercepts the y axis.

The regression line in Box 9.6 was calculated so that the error sum of squares (SS) was minimized, i.e. least squares. Analysis of variance (ANOVA) is used to calculate how much of the total SS is due to the regression and how much is residual (noise within the system, unexplained variation). The null hypothesis is that the regression does not explain a significant proportion of the variation in y-values, i.e. has slope zero. In the example in Box 9.6, $R^2 = 0.95$, which means that 95% of the variability in the data can be explained by the fitted line.

9.5.2 SPEARMAN'S RANK

Like the Mann–Whitney test, Spearman's rank correlation is calculated from ranked data, but this time each variable is ranked separately in ascending order. In Box 9.6 the regression line is significantly different from zero, i.e. it has a significantly negative slope. Spearman's rank correlation, calculated in the usual way, has a value of -1.

9.6 EFFECTS ASSESSMENT

The biological effects caused by substances of concern are also examined in the second phase of a predictive risk assessment. This is often referred to as an effects assessment and is normally achieved through a series of toxicity tests or bioassays. Statistical analyses are essential tools for interpreting the outcomes of a bioassay (Crane and Chapman 1996). This is because bioassays use individual organisms, each with a different tolerance to toxic chemicals, which produce a distribution of responses from the most sensitive to the least sensitive. This 'statistical tolerance distribution' applies both when data are continuous (such as measurements of growth) and when data are quantal (all-or-nothing responses such as survival). The standard procedure for single species bioassays is to select a concentration series of the test substance and expose a separate group of test organisms to each of these. Recordings are made of the response (usually mortality, or reductions in growth or reproduction) at each concentration (Forbes 1993). This response is then used to estimate the concentration–response curve and the point estimates currently required by risk assessors, such as the median lethal concentration (LC_{50}) and the no observed effect concentration (NOEC). Each of these is usually estimated after a specific exposure duration. For example, estimates from acute ecotoxicity tests with the crustacean *Daphnia magna* are usually reported as a 48 h LC_{50}, while chronic fish tests normally last for several days or weeks and report an effective concentration (EC) value and a NOEC.

The response of test organisms to toxicants depends not only upon the dose or concentration to which they are exposed, but also on the duration of that exposure. Risk assessors recognise this, but formal statistical methods for incorporating exposure duration currently tend to be eschewed in favour of qualitative graphical approaches (Suter *et al.* 1987), the derivation of acute-to-chronic ratios (Kenaga 1982), or the use of application factors (Mount and Stephan 1967). If the time course of toxicity could be taken into account it may be possible to make predictions of chronic toxicity on the basis of acute results, improve the precision of LC_{50} estimates and estimate true no effect concentrations (NEC) instead of the standard no observed effect concentration (NOEC), which is considered as non-protective by many statisticians (Chapman *et al.* 1996; OECD 1997). Newman (1995) also argues that if data on the time course of effects on survival and reproduction are available, ecologically meaningful endpoints such as the intrinsic rate of population increase can also be estimated.

Several approaches for the estimation of NECs and/or acute:chronic extrapolations have been developed recently. Mayer *et al.* (1991, 1994) have used two-step linear regression and multifactor probit analysis. Dixon and Newman (1991), Newman and Aplin (1992) and Sun *et al.* (1995) have recommended the use of survival time modelling and accelerated life testing, and Bedaux and Kooijman (1994) have proposed theoretically derived functions to take explicit account of the time dependence of toxicity. These approaches represent a theoretical improvement on the standard methods, but have not yet been widely used in risk assessment. In this section we will illustrate the use of several of these standard and novel statistical techniques to analyse a simple data set from a fish toxicity test.

9.6.1 ESTIMATION OF A NOEC USING ANOVA AND MULTIPLE COMPARISON TESTS

The NOEC can be calculated from concentration–response data by using analysis of variance (ANOVA), followed by a multiple comparison test, so long as there is replication at each concentration. A simple example for our substance is shown in Box 9.7. Ten juvenile fish are assigned at random among three replicates, each of four concentrations of the substance and a control. They are left in the water for 96 h and observed at 24 h intervals to see whether they are still alive. 'Concentration' is the factor of interest, and it has five 'levels' (0, 1, 10,

---

Box 9.7 Calculation of a NOEC using ANOVA and a multiple comparison test

Survival of fish in three replicates (10 fish per replicate)

	0 mg l^{-1}	5 mg l^{-1}	10 mg l^{-1}	15 mg l^{-1}	20 mg l^{-1}
	10	8	5	3	0
	9	7	4	2	1
	10	6	5	1	0

Survival of fish in three replicates: arcsin transformed

	0 mg l^{-1}	5 mg l^{-1}	10 mg l^{-1}	15 mg l^{-1}	20 mg l^{-1}
	1.57	1.11	0.79	0.58	0.00
	1.25	0.99	0.68	0.46	0.32
	1.57	0.89	0.79	0.32	0.00
Mean	1.46	0.99	0.75	0.46	0.11

Analysis of Variance (Balanced Designs)

```
Factor   Type  Levels Values
conc     fixed    5      0     5     10     15     20
```

Analysis of Variance

```
Source     DF    SS        MS        F        P
conc        4    3.20736   0.80184   39.58    0.000
Error      10    0.20261   0.02026
Total      14    3.40997
```

There is a significant difference in mortality between the different test concentrations.

Tukey test

```
Conc   mean¹
  0    1.46  a
  5    0.99  b
 10    0.75  bc
 15    0.46  cd
 20    0.11  d
```

[1]On transformed scale.

Means, with no following letters in common, are significantly different from one another at $p = 0.05$. Many of these concentrations produce effects that differ significantly from each other. However, a NOEC cannot be determined with this Tukey test because the lowest concentration produced a response that was significantly different to the control.

Continued overleaf

Other multiple comparison tests

Toxstat 3.1 (West Inc. 1994) was used to estimate a NOEC using several different multiple comparison tests.

Dunnett's test	NOEC <5 mg l^{-1}
Bonferroni t-test	NOEC <5 mg l^{-1}
Williams test	NOEC <5 mg l^{-1}
Wilcoxon rank sum test with Bonferroni adjustment	NOEC <5 mg l^{-1}
Dunn's multiple comparison	NOEC $= 15$ mg l^{-1}

Most of the results showed that the NOEC was below the lowest test concentration. However, Dunn's multiple comparison, a non-parametric test with low power, produced a NOEC of 15 mg l^{-1} because only the 20 mg l^{-1} treatment differed significantly from the control.

---

15 and 20 mg l^{-1}). Proportion surviving (arcsin transformed) is the dependent variable. We want to know whether fish survival is the same after 96 h exposure to the five different concentrations. ANOVA will provide a probabilistic answer to this question. ANOVA makes two main assumptions: (1) the data in each sample are drawn from populations of normally distributed, continuous variables; (2) the variances for each group are equal. Like most common parametric tests, ANOVA is said to be robust to deviations from these assumptions. But if there are major departures from normality, the data should either be transformed or the non-parametric Kruskal–Wallis test used.

So, by using ANOVA we are able to make a probability statement about whether different samples are likely to come from the same population. In the above example we can say that fish survival after 96 h differs depending upon the concentration of substance to which they were exposed. However, after this we will usually want to know which of the groups differ significantly from each other. We can do this by using a multiple comparison test. A parametric example of this is the Tukey test (see Chapter 3). There are many different parametric multiple comparison tests, including the Williams test, the Bonferroni t-test and Dunnett's test. The choice depends mainly on the precise nature of the comparisons. Non-parametric alternatives are also available for use after a Kruskal–Wallis test.

The perceived advantage of the NOEC is that it is easy to understand (OECD 1997). However, there are many disadvantages to its use (Chapman *et al.* 1996). These disadvantages are discussed in Chapter 4. Problems with the NOEC have led many statisticians and biologists to propose that hypothesis testing is not well suited to the type of data obtained from most ecotoxicity tests (Chapman *et al.* 1996), except in the special case of limit tests (Whitehouse *et al.* 1996). Box 9.7 gives an example of a NOEC calculation for our substance. Clearly, this NOEC is highly dependent upon the multiple range test used in its calculation. If the parametric tests are used, no NOEC can be calculated because the lowest test concentration causes a statistically significant effect. If the non-parametric alternative is used, the NOEC is 15 mg l^{-1}.

9.6.2   DOSE RESPONSE ANALYSIS

The estimation of an EC_x value (the EC at a specified value of x – usually EC_{50}) overcomes most of the problems associated with hypothesis testing (Chapman *et al.* 1996), and is the usual form of analysis for acute ecotoxicity experiments, as discussed in Chapter 4.

Data from fixed times of observation (usually 24, 48, 72 or 96 h in acute bioassays) are transformed so that least squares fits can be made to linear models. Linear models have advantages, despite the availability of non-linear curve-fitting routines in most statistical software packages. The estimation of confidence limits is easier and methods for checking model fit are better developed for linear models (Forbes 1993). Linearity is usually achieved by logging the exposure concentration and converting the response to its probit (Bliss 1935; Finney 1971) or logit (Berkson 1944). Several other linearising transformations are also available, such as the arcsine or Weibull (Weibull 1951). More recently, Generalised Linear Models have been proposed as a method of analysing data without the need for an initial linearising transformation (Kerr and Meador 1996).

Whatever the derivation of the dose–response curve, EC_x values are then estimated for the magnitude of effect that interests the investigator. This is normally an EC_{50} or LC_{50}, because more precise estimates are usually possible at this median point. However, the x in EC_x can be as large or small as an investigator wishes, although estimates at the extremes of the probability function are likely to have very wide confidence intervals. Hartley and Sielken (1977) summed up this problem as follows:

Experiments attempting to measure . . . minute differential risk increments *directly* by using only extremely small . . . *residual doses* are forced, in the face of statistical errors, to use astronomically large numbers of animals. On the other hand, if experiments are conducted at adequately high doses (accelerated doses) the problem of extrapolating (or interpolating) . . . to residual dose levels arises.

Bruce and Versteeg (1992) discussed the choice of x in EC_x and concluded that a value of 20% is normally protective when the natural variability of populations is taken into account. However, many researchers would consider a value of 20% effect as too high (OECD 1997). Furthermore, the choice of different EC_x values often leads to differences in the toxicity ranking of samples if the response slopes are not parallel (Oris and Bailer 1997).

Partly because of problems in deciding what x should be estimated in EC_x, much effort has been expended on the estimation of 'thresholds' below which no toxic effects occur (Cox 1987). The problem with thresholds is that they are, like other concentration-response models, highly model dependent. The estimates of thresholds from different models may vary widely, and the data from standard ecotoxicity experiments provide little information on which model is correct (Hoekstra and van Ewijk 1993). To avoid these apparent shortcomings in the estimation of a 'safe' threshold, Crump (1984) proposed the estimation of 'benchmark' concentrations and Hoekstra and van Ewijk (1993) proposed 'bounded effect' concentrations. These are approaches that do not rely so heavily on model assumptions.

Bounded-effect concentrations are estimated by first selecting a concentration (the bounded-effect concentration) for linear extrapolation to a lower concentration. Hoekstra and van Ewijk (1993) suggest that the concentration corresponding to ≤25% effect is often a justifiable choice for the bounded effect concentration. If 25% effect is chosen as the criterion, then the bounded effect concentration should be the highest *tested* concentration for which the upper confidence limit does not exceed 25% effect. Next, the concentration at which an acceptably small effect occurs (e.g. 1% effect) is estimated by linear extrapolation from the bounded effect concentration.

Box 9.8 shows how the fish data in Box 9.7 can be analysed using different dose response models to produce LC values and a bounded effect concentration.

Box 9.8 Time-dependent dose response analysis of fish toxicity data

Concentration (mg l^{-1})	Log concentration	No. dead/ total dead	Proportion dead	Arcsin	Probit	Logit	Weibull
0	—	1/30	0.03	0.17	3.12	−3.48	−3.49
5	0.699	9/30	0.30	0.58	4.48	−0.85	−1.03
10	1.0	16/30	0.53	0.82	5.08	0.12	−0.28
15	1.176	24/30	0.80	1.11	5.84	1.39	0.48
20	1.301	29/30	0.97	1.40	6.88	3.48	1.25

Graphical approach
A graphical method can be used to analyse these data (Litchfield and Wilcoxon 1949) and produce estimates of LC_x and associated confidence intervals. Figure 9.6 shows that estimates of LC_{50} do not differ substantially using any of the usual transformations of effect.

Confidence limits for the probit estimate of LC_{50} are calculated as follows:
From the graph, $LC_{16} = 2.8$, $LC_{50} = 9.0$ and $LC_{84} = 15.0$

$$\text{Slope function } (S) = \frac{(LC_{84}/LC_{50})(LC_{50}/LC_{16})}{2} = \frac{(15/9) + (9/28)}{2} = \frac{1.7 + 3.2}{2} = 2.45$$

Number of individuals tested between 16% and 84% responses $(N') = 90$

$$f_{LC_{50}} = S^{2.77/\sqrt{N'}} = 2.45^{2.77/\sqrt{90}} = 1.3$$

95% CI for the $LC_{50} = 9.0 \pm 1.3$ mg l^{-1}

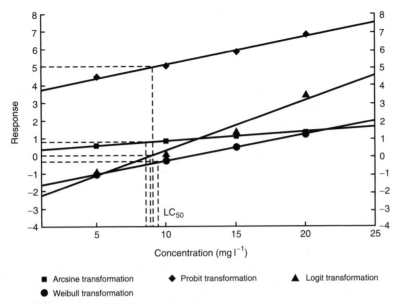

Figure 9.6   Concentration-response using different response transformations.

Computer-based approach
The graphical approach has now been largely superseded by computer-based
maximum likelihood methods (Newman 1995) using software such as Toxstat 3.4
(West Inc. 1994). The chi-square values calculated for goodness of model fit suggest
that the probit is slightly better than the logit. However, there is little difference in
the estimates of LC_x by the two parametric methods, or the non-parametric
Spearman–Karber method.

Probit analysis
96 h LC_{50} = 8.05 mg l^{-1} (95% CI   (6.82, 9.50))
96 h LC_{20} = 4.59 mg l^{-1} (95% CI   (3.62, 5.84))
96 h LC_{10} = 3.43 mg l^{-1} (95% CI   (2.54, 4.63))
χ^2 lack of fit = 3.011

Logit analysis
96 h LC_{50} = 8.14 mg l^{-1} (95% CI   (6.84, 9.68))
96 h LC_{20} = 4.71 mg l^{-1} (95% CI   (3.64, 6.09))
96 h LC_{10} = 3.41 mg l^{-1} (95% CI   (2.44, 4.77))
χ^2 lack of fit = 3.626

Spearman–Karber analysis
96 h LC_{50} = 7.41 mg l^{-1} (95% CI   (5.74, 9.57))

Bounded effect concentration
The lowest tested concentration (5 mg l^{-1}) has an upper confidence limit of 41%
effect. This is rather large, but is the best that is available with this dataset. The 1%
bounded effect is then calculated as $x = 5/41 = 0.12$ mg l^{-1}.

9.6.3 ANALYSES INCLUDING EXPOSURE DURATION

Although there is general agreement that use of a dose–response curve to estimate an EC_x or a threshold, benchmark or bounded effect concentration has many advantages over the derivation of a NOEC, the calculation of an EC_x at specific time intervals still uses data sub-optimally. This is because most investigators will take some measurements during the course of a bioassay, especially if survival is the endpoint. These data from intermediate observation periods are usually not reported or used in the final estimation.

Traditionally, the standard analysis of exposure duration was to plot the mean or median survival time (or its reciprocal) against the toxicant concentration or log concentration (Sprague 1969). A line was then fitted either by eye, or by using a more formal model (e.g. Hey and Hey 1960). A 'safe' level of chemical or effluent over an indefinite period was then estimated from this (e.g. Alderdice and Brett 1957).

More recently, Mayer *et al.* (1991, 1994) developed a simple statistical approach for using observations at different time periods in acute tests to predict chronic lethality. They made two assumptions essential for any prediction of this type: (1) the concentration response is a continuum in time; (2) the mode of action and detoxification systems for lethality are similar under acute and chronic exposures. Linear regression is used to estimate an LC_0 at all observation times (Probit percentage mortality $= a + b(\log$ concentration)). These estimates of LC_0 are then regressed against the reciprocal of time ($LC_0 = a + b(1/t)$). The intercept of the regression line is the chronic 'predicted no observed effect concentration' (PNOEC).

Sun *et al.* (1995), in a continuation of the work described above, have recommended the use of survival time modelling and accelerated life testing. This has an advantage over the earlier work by Mayer *et al.* (1991, 1994), in that the statistical dependence of observations at different times does not present analytical difficulties. Survival time models have been proposed by other researchers as a method for integrating time, concentration and response, and covariables such as organism weight (Dixon and Newman 1991; Newman and Aplin 1992; Newman *et al.* 1989, 1994). These models belong to the family of accelerated failure time models, commonly used by engineers and medical statisticians (Kalbfleisch *et al.* 1983). Sun *et al.* (1995) have described a model in which organism survival at a particular period of time in a particular concentration depends upon the strength of toxic action, the concentration–response surface shape and the time-response surface shape. A software package called ACE (acute to chronic estimation) is available for performing these calculations (Mayer *et al.* 1996).

Several authors have proposed the use of simple kinetic models to overcome the problem of determining NEC values, EC_x at infinite exposure time, or both of these (e.g. Chen and Selleck 1969; Chew and Hamilton 1985; Heming *et al.* 1989; Matida 1960). Compared with the empirical models described above, these kinetic models involve more biological assumptions about the behaviour of toxicants in living organisms (Kalbfleisch *et al.* 1983). In the most recent of these attempts, Kooijman (1993, 1996) uses hazard-based models that replace estimates of the LC_x and the slope of the response with a single parameter, the killing rate (defined as the probability of dying, per unit of time and per unit of environmental concentration that exceeds the NEC (Bedaux and Kooijman 1994). The mechanistic basis of this model allows the estimation of toxic effects on reproduction and growth in different species using the same approach (Kooijman and Bedaux 1996a, b, c; Kooijman *et al.* 1996). A further advantage is that it is possible to model the effects of episodic exposure to toxicants and the effect of non-persistent chemicals that degrade relatively rapidly (Widianarko and van Straalen 1996). A software package, Debtox, is available for making the calculations described above (Kooijman and Bedaux 1996d).

Box 9.9 provides summaries from analyses of the fish data using the ACE and Debtox software packages. The different models produce quite different estimates of 'safe' levels of our substance. Two-step linear regression predicts that $0.019 - 0.452$ mg l^{-1} will cause only a negligible effect of 0.01% after indefinite exposure. The accelerated life testing model predicts a 0.01% effect at between

---

Box 9.9   Two-step linear regression, accelerated life testing and hazard model analysis of toxicity data

Dataset

			Survival		
Concentration	0 h	24 h	48 h	72 h	96 h
0	30	29	29	29	29
5	30	29	24	22	21
10	30	26	21	17	14
15	30	19	10	8	6
20	30	15	1	1	1

Two-step linear regression (ACE)

Form of model

1. Probit mortality $= a + b$ (log concentration)
2. $LC_0 = a + b$ (1/time)

Model results

Effect level (%)	Time independent LC_x (mg l^{-1})	95% CI (mg l^{-1})	R^2
0.01	0.092	0.019–0.452	0.95
10	0.417	0.350–0.496	0.99
20	3.37	3.16–3.6	0.99
50	5.85	5.22–6.54	0.99

Continued overleaf

Accelerated life testing (ACE) *hazard model*

Form of model

$$q(t; c) = \exp(-ac^b t^d)$$

where $q(t; c)$ is the survival probability at time t and environmental concentration c, a the measure of strength of toxic action, b the dose-response surface shape and d the time-response surface shape.

Model results

Effect level (%)	90 day LC_x (mg l^{-1})	95% CI (mg l^{-1})
0.01	0.038	(0.008, 0.087)
5	0.650	(0.130, 1.17)

Debtox hazard model

Form of model

$$q(t; c) = \exp\left(\frac{\dot{k}_\dagger}{\dot{k}_a} c(e^{-\dot{k}_a t_0} - e^{-\dot{k}_a t}) - \dot{k}_\dagger (c - c_0)(t - t_0)\right) \quad \text{if } c > C_0 \text{ and } t > t_0$$

where $q(t; c)$ is the survival probability at time t and environmental concentration c, $\dot{k}_\dagger$ the killing rate and $\dot{k}_a$ the elimination rate of chemical from organism.

Model results

96 h LC_{50} = 7.29 mg l^{-1} (± 0.56)
96 h LC_{20} = 4.59 mg l^{-1} (± 0.62)
96 h LC_{10} = 3.9 mg l^{-1} (± 0.67)
Time independent NEC = 3.25 mg l^{-1} (95% CI (1.26, 4.15))

0.008 and 0.087 mg l^{-1}, while the hazard model in Debtox predicts a safe concentration at between 1.26 and 4.15 mg l^{-1}. Clearly, further research is necessary before any of these models can be used routinely in risk assessment.

9.7 RISK CHARACTERIZATION AND MANAGEMENT

The final two phases in an ecological risk assessment draw together the information gathered in earlier phases and allow the risk assessor and risk manager to come to an appropriate decision. During the risk characterization phase the assessor will take account of the concentration of the hazardous substance, the duration of exposure, the proportion of individuals, populations or assemblages responding, and the severity of these effects. During the risk management phase an attempt will be made to minimize risks without undue harm to other societal values.

An assessor could characterize the risks posed by a substance by taking the most conservative values available for exposure (e.g. predicted worst case maximum exposure over a prolonged period) and toxic effect (e.g. chronic sublethal effects on the most sensitive species). This may be sufficient if the predicted environmental concentration (PEC) and the predicted no effect concentration (PNEC) obtained from this ultraconservative approach suggest that environmental concentrations of the substance will be substantially below levels that cause biological effects, even when safety factors are applied. This simple toxicity: exposure ratio (TER) approach can be used to screen out substances that are probably 'safe' in the environment (Campbell and Hoy 1996). However, the TER approach to assessment may be overprotective for some substances, leading to unnecessary costs for both dischargers and regulators (Whitehouse and Cartwright, 1998).

Several aspects of toxicity and exposure need to be considered if a risk assessor wishes to estimate concentrations that are just sufficient to protect environmental values. The sensitivity of the overall assessment to different components in the assessment model should be determined as should the level of uncertainty in the estimation of each of these components. Finally, an assessment of ecological risk should be estimated by combining the probability of exposure with the probability of harm. In this section we show how to perform simple sensitivity and uncertainty analyses and how to calculate the probability of harm. We have insufficient space to deal with the complex issue of risk management, although there have been some exciting recent advances in the use of decision analysis in chemical risk assessment.

9.7.1   SENSITIVITY AND UNCERTAINTY ANALYSES

Sensitivity analysis can be used to examine a wide range of different model inputs to determine which contribute most to uncertainty in model outputs (Morgan and Henrion 1990). This is important during risk characterization because an ecological risk assessor will often consider only a few dimensions. For example, duration of exposure is often held constant in toxicity assessments, and emphasis placed upon exposure concentration (Suter 1995). Sensitivity analysis can help to determine which dimensions should be included. Uncertainty analysis can be used to estimate the uncertainty in model output by quantifying uncertainties among the inputs (Dakins et al. 1994). Monte Carlo simulation is the most frequently used method for quantifying uncertainty (Barnett and O'Hagan 1997), and user-friendly packages such as Crystal Ball 4.0 (Sargent and Wainwright 1996) are available. In a Monte Carlo simulation the assessor chooses a probability density function, such as a normal or exponential curve, to describe the distribution of the measurement endpoints. A large number of simulations are then run with different input values to generate many output values from which percentiles can be calculated.

Box 9.10 shows, in very simplified form, how sensitivity analysis can be used to identify components of an equation with the most important effects on the final calculated value. In this example we take the two-step linear regression model used in Box 9.9 and identify whether the slope of the response or the estimate of mean LC_0 has the greatest influence on the predicted 'safe' level. The results show that estimates of the mean LC_0 have a considerably greater influence. Most uses of sensitivity analysis are more complex than this example, but the principles remain the same.

Box 9.10 also shows the results from a Monte Carlo analysis of $LC_{0.001}$ values. An exponential distribution was assumed (mostly low, but some high LC values) with a mean and SD based upon the values predicted in the sensitivity analysis above. The programme was allowed to run for 1000 iterations. The results suggest that with these assumptions the $LC_{0.001}$ could range from 0 to 0.742 mg l^{-1}. This information is useful to a risk manager because it allows them to determine whether gaining more information or reducing uncertainty in some variables could produce an improved decision (Dakins *et al.* 1994).

Box 9.10   Sensitivity and uncertainty analysis during risk characterization

Sensitivity analysis for toxicity to fish
Realistic low, medium and high estimates of the regression slope coefficient and the mean intercept (LC_0) were calculated from the data in Box 9.9. The values in the matrix are the predicted $LC_{0.001}$ at infinite time. At high slope values there is a sevenfold difference in the predicted $LC_{0.001}$.

	LC_0 (mg l^{-1})		
	Low (0.15)	Medium (0.19)	High (0.53)
Slope (b)			
Low (0.613)	0.136	0.176	0.52
Medium (2.21)	0.10	0.139	0.48
High (3.85)	0.06	0.101	0.44

Uncertainty analysis

Percentiles	Distribution of values for potential $LC_{0.001}$ at infinite time
10	−0.024
20	0.059
30	0.112
40	0.161
50	0.217
60	0.256
70	0.304
80	0.352
90	0.423
100	0.742

9.7.2   COMBINING EXPOSURE AND EFFECTS ASSESSMENTS

The last task of the risk assessor before the risk management phase is to combine estimates of exposure and effect. This can done probabilistically in a
simple graphical fashion as shown in Box 9.11. In this example we have taken
exposure data based on those reported in Box 9.2 and the toxicity data for fish
reported in Box 9.7. The approach, illustrated for the sake of simplicity by using
just one organism and one estimate of exposure, can theoretically be extended
to take into account many species and exposure scenarios (e.g. Aldenburg and
Slob 1991; Solomon *et al.* 1996; Novotny and Witte 1997). However, some of
the more complex models may contain either untestable or untenable assumptions (Forbes and Forbes 1993). This is currently an area of intense research
activity by ecotoxicologists and statisticians.

Box 9.11   Combining exposure and effects estimates to calculate risk

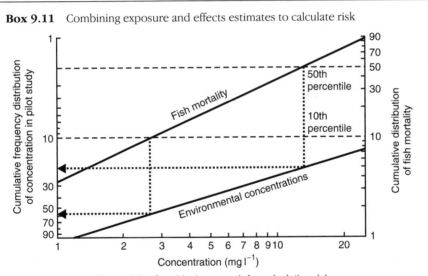

Figure 9.7   Graphical approach for calculating risk.

The figure above is a simple graphical approach to the calculation of risk. Fish
mortality (or any other effects assessment) is plotted as a cumulative frequency
distribution. Estimated environmental concentrations (or any other exposure
assessment) are also plotted as a cumulative frequency. To estimate the probability
that an environmental concentration will occur that is sufficient to kill > 50% of this
fish species, a vertical line is drawn from the intersection of the 50th percentile for
fish mortality and the fish mortality trend line. When this line meets the trend line
for environmental concentrations, a horizontal line is drawn until it intersects the
concentration axis. The point of intersection provides an estimate of the probability
of occurrence. In this case, > 10 % are likely to be killed approximately 55% of the
time given this distribution of environmental concentrations. Solomon *et al.* (1996)
provide an interesting example of this approach to risk characterization.

9.8 CONCLUSIONS

In this chapter we have tried to show the breadth of statistical techniques available to chemical risk assessors, from simple summary statistics to more complex geostatistical and modelling techniques. We have emphasized the importance of clearly specifying the aims of the assessment, preferably within the well-tested framework already provided by decision analysis. We have provided a very quick tour around some of the more frequently used approaches, plus one or two more novel methods.

All of the techniques that we have described are implemented in readily available software. However, caution should be exercised by new users. It is all too easy to generate apparently plausible results from faulty assumptions when one has gigabytes of computing power at one's disposal. A professional statistician should always be consulted before a decision about chemical safety in the environment is made on the basis of statistical analysis. Our view is that the role and duty of the environmental scientist is to learn enough about statistics to hold an intelligent conversation with a statistician.

ACKNOWLEDGEMENT

Part of Mark Crane's contribution to this chapter was funded by the Environment Agency contract EMA 003.

REFERENCES

Aldenburg T and Slob W (1991) Confidence limits for hazardous concentrations based on logistically distributed NOEC toxicity data. *Ecotoxicology and Environmental Safety*, **25**, 48–63.

Alderdice DF and Brett JR (1957) Some effects of kraft mill effluent on young Pacific salmon. *Journal of the Fisheries Research Board of Canada*, **14**, 783–795.

Barnett V and O'Hagan T (1997) *Setting Environmental Standards: The Statistical Approach to Handling Uncertainty and Variation.* Report to the Royal Commission on Environmental Pollution.

Bedaux JJM and Kooijman SALM (1994) Statistical analysis of bioassays based on hazard modeling. *Journal of Environmental Statistics*, **1**, 303–314.

Berkson J (1944) Application of the logistic function to bioassay. *Journal of the American Statistical Society*, **39**, 357–365.

Bliss CI (1935) The calculation of the dosage-mortality curve. *Annals of Applied Biology*, **22**, 134–167.

Bruce RD and Versteeg DJ (1992) A statistical procedure for modeling continuous toxicity data. *Environmental Toxicology and Chemistry*, **11**, 1485–1494.

Calow P (1992) Can ecosystems be healthy? Critical consideration of concepts. *Journal of Aquatic Ecosystem Health*, **1**, 1–5.

Campbell PJ and Hoy SP (1996) ED points and NOELs: how they are used by UK regulators. *Ecotoxicology*, **5**, 139–144.

Chapman PF, Crane M, Wiles J, Noppert F and McIndoe E (1996) Improving the quality of statistics in regulatory ecotoxicity tests. *Ecotoxicology*, **5**, 169–186.

Chen CW and Selleck RE (1969) A kinetic model of fish toxicity threshold. *Journal of the Water Pollution Control Federation*, **41**, 294–308.

Chew RD and Hamilton MA (1985) Toxicity curve estimation: fitting a compartment model to median survival times. *Transactions of the American Fisheries Society*, **114**, 403–412.

Cowan CE, Mackay D, Feijtel TCJ, van de Meent D, di Guardo A, Davies J and Mackay N (1995) *The Multi-Media Fate Model: A Vital Tool for Predicting the Fate of Chemicals*, SETAC Press, Pensacola, FL.

Cox C (1987) Threshold dose-response models in toxicology. *Biometrics*, **43**, 511–524.

Crane M and Chapman PF (1996) Asking the right questions: ecotoxicology and statistics. *Ecotoxicology*, **5**, 137–138.

Crane M and Newman MC (1996) Scientific method in environmental toxicology. *Environmental Reviews*, **4**, 112–122.

Cressie NAC (1990) The origins of kriging. *Mathematical Geology*, **22**, 239–252.

Crump KS (1984) A new method for determining allowable daily intakes. *Fundamental Applications of Toxicology*, **4**, 854–871.

Dakins ME, Toll JE and Small MJ (1994) Risk-based environmental remediation: decision framework and role of uncertainty. *Environmental Toxicology and Chemistry*, **13**, 1907–1915.

Davies BE (1989) Data handling and pattern recognition for metal contaminated soils. *Environmental Geochemistry and Health*, **11**, 137–143.

Dixon PM and Newman MC (1991) Analysing toxicity data using statistical models of time-to-death: an introduction. In *Metal Ecotoxicology: Concepts and Applications*, Newman MC and McIntosh AW (eds), Lewis Publishers, Chelsea, MI, pp. 207–242.

DoE (1994) *Sampling Strategies for Contaminated Land*, Department of the Environment Contaminated Land Research Report No. 4. Nottingham Trent University.

European Commission (1996) *Technical Guidance Document in Support of Commission Directive 93/67/EEC on Risk Assessment for New Notified Substances and Commission Regulation (EC) No 1488/94 on Risk Assessment for Existing Substances. Part II Environmental Risk Assessment*, Office for Official Publications of the European Community, Luxembourg.

Finney DJ (1971) *Probit Analysis*, 3rd edn, Cambridge University Press, Cambridge, UK.

Flatman GT and Yfantis AA (1996) Geostatistical sampling designs for hazardous waste sites. In *Principles of Environmental Sampling*, 2nd edn, Keith LH (ed.), American Chemical Society, Washington, DC pp. 779–801.

Forbes TL (1993) The design and analysis of concentration-response experiments. In *Handbook of Ecotoxicology*, Volume 1, Calow, Peter (ed), Blackwell Science, Oxford, UK. pp. 438–459.

Forbes TL and Forbes VE (1993) A critique of the use of distribution-based extrapolation models in ecotoxicology. *Functional Ecology*, **7**, 249–259

Fox GA (1991) Practical causal inference for ecoepidemiologists. *Journal of Toxicology and Environmental Health*, **33**, 359–373.

Gilbertson M, Kubiak T, Ludwig J and Fox G (1991) Great Lakes embryo mortality, edema, and deformities syndrome (GLEMEDS) in colonial fish-eating birds: similarity to chick-edema disease. *Journal of Toxicology and Environmental Health*, **33**, 455–520.

Ginevan ME and Splitstone DE (1997) Improving remediation decisions at hazardous waste sites with risk-based geostatistical analysis. *Environmental Science and Technology* **31**, 92–96.

Golden Software Inc. (1997) *Surfer for Windows Version 6. User's Guide*. Golden Software, Golden, Colorado.

Hartley HO and Sielken RL Jr (1977) Estimation of 'safe doses' in carcinogenic experiments. *Biometrics*, **33**, 1–30.

Heming TA, Sharma A and Kumar Y (1989) Time–toxicity relationships in fish exposed to the organochlorine pesticide methoxychlor. *Environmental Toxicology and Chemistry*, **8**, 923–932.

Hey EN and Hey MH (1960) The statistical estimation of a rectangular hyperbola. *Biometrics*, **16**, 606–617.

Hoekstra JA and van Ewijk PH (1993) Alternatives for the no-observed effect level. *Environmental Toxicology and Chemistry*, **12**, 187–194.

Isaaks EH and Srivastava RM (1990) *An Introduction to Applied Geostatistics*, Oxford University Press, NY.

Journel AG (1986) *Geostatistics for the Environmental Sciences*, Stanford University Press, CA.

Kalbfleisch JD, Krewski DR and van Ryzin J (1983) Dose–response models for time-to-response toxicity data, *Canadian Journal of Statistics*, **11**, 25–49.

Kenaga EE (1982) Predictability of chronic toxicity from acute toxicity of chemicals in fish and aquatic invertebrates. *Environmental Toxicology and Chemistry*, **1**, 347–358.

Kerr DR and Meador JP (1996) Modeling dose response using generalized linear models. *Environmental Toxicology and Chemistry*, **15**, 395–401.

Kooijman SALM (1993) *Dynamic Energy Budgets in Biological Systems: Theory and Applications in Ecotoxicology*. Cambridge University Press, Cambridge.

Kooijman SALM (1996) An alternative for NOEC exists, but the standard model has to be abandoned first. *Oikos*, **75**, 310–316.

Kooijman SALM and Bedaux JJM (1996a) Analysis of toxicity tests on fish growth. *Water Research*, **30**, 1633–1644.

Kooijman SALM and Bedaux JJM (1996b) Analysis of toxicity tests on *Daphnia* survival and reproduction. *Water Research*, **30**, 1711–1723.

Kooijman SALM and Bedaux JJM (1996c) Some statistical properties of estimates of no-effect concentrations. *Water Research*, **30**, 1724–1728.

Kooijman SALM and Bedaux JJM (1996d) *The Analysis of Aquatic Toxicity Data*. VU University Press, Amsterdam, 149 pp.

Kooijman SALM, Hanstveit AO and Nyholm N (1996) No-effect concentrations in algal growth inhibition tests. *Water Research* **30**, 1625–1632.

Lichfield JT Jr and Wilcoxon F (1949) A simplified method of evaluation dose-effect experiments. *Journal of Pharmacology and Experimental Therapeutics*, **96**, 99–113.

Mackay D (1991) *Multimedia Environmental Fate Models: The Fugacity Approach*, Lewis Publishers, Chelsea, MI.

Maltby L and Calow P (1989) The application of bioassays in the resolution of environmental problems; past, present and future. *Hydrobiologia*, **188/189**, 65–76.

Matida Y (1960) Study on the toxicity of agricultural control chemicals in relation to fisheries management. No. 3. A method for estimating threshold value and a kinetic analysis of the toxicity curve. *Bulletin of the Freshwater Fisheries Research Laboratory, Tokyo*, **9**, 1–12.

Mayer FL, Krause GF, Ellersieck MR and Lee G (1991) Statistical approach to predicting chronic toxicity of chemicals to fishes from acute toxicity test data. PB92-169655, National Technical Information Service, Springfield, VA.

Mayer FL, Krause GF, Buckler DR, Ellersieck MR and Lee G (1994) Predicting chronic lethality of chemicals to fishes from acute toxicity test data: concepts and linear regression analysis. *Environmental Toxicology and Chemistry*, **13**, 671–678.

Mayer FL, Sun K, Lee G, Ellersieck MR and Krause GF (1996) *User Guide: Acute to Chronic Estimation (ACE)*, United States Environmental Protection Agency, Gulf Breeze, FL.

Morgan GM and Henrion M (1990) *Uncertainty: A Guide to Dealing with Uncertainty in Quantitative Risk and Policy Analysis*, Cambridge University Press, Cambridge.

Mount DI and Stephan CE (1967) A method for establishing acceptable limits for fish – malathion and the butoxyethanol ester of 2,4-D. *Transactions of the American Fisheries Society*, **96**, 185–193.

National Research Council (1993) *Issues in Risk Assessment:* Volume 3, *Ecological Risk Assessment*, National Academy Press, Washington, DC.

Newman MC (1995) *Quantitative Methods in Aquatic Ecotoxicology*, Lewis Publishers, Boca Raton, FL.

Newman MC and Aplin M (1992) Enhancing toxicity data interpretation and prediction of ecological risk with survival time modeling: an illustration using sodium chloride toxicity to mosquitofish (*Gambusia holbrooki*). *Aquatic Toxicology*, **23**, 85–96.

Newman MC, Diamond SA, Mulvey M and Dixon P (1989) Allozyme genotype and time to death of mosquitofish, *Gambusia affinis* (Baird and Girard) during acute toxicant exposure: a comparison of arsenate and inorganic mercury. *Aquatic Toxicology*, **15**, 141–156.

Newman MC, Keklak MM and Doggett MS (1994) Quantifying animal size effects on toxicity: a general approach. *Aquatic Toxicology*, **28**, 1–12.

Novotny V and Witte JW (1997) Ascertaining aquatic ecological risks of urban stormwater discharges. *Water Research*, **31**, 2573–2585.

OECD (1997) *Draft Report of the OECD Workshop on Statistical Analysis of Aquatic Toxicity Data, Braunschweig, Germany, 15–17 October 1996*, Organization for Economic Cooperation and Development, Paris, 50 pp.

Oris JT and Bailer AJ (1997) Equivalence of concentration–response relationships in aquatic toxicology studies: testing and implications for potency estimation. *Environmental Toxicology and Chemistry*, **16**, 2204–2209.

Power M and McCarty LS (1997) Fallacies in ecological risk assessment practices. *Environmental Science and Technlogy*, **31**, 370–375.

Reckhow KH (1994) A decision analytic framework for environmental analysis and simulation modeling. *Environmental Toxicology and Chemistry*, **13**, 1901–1906.

Sargent R and Wainwright E (1996) *Crystal Ball Version 4.0 User Manual*, Decisioneering Inc. , Denver, CO.

Solomon KR, Baker DB, Richard RP, Dixon KR, Klaine SJ, La Point TW, Kendall RJ, Weisskopf CP, Giddings JM, Giesy JP, Hall JW Jr and Williams WM (1996) Ecological risk assessment of atrazine in North American surface waters. *Environmental Toxicology and Chemistry*, **15**, 31–76.

Sprague JB (1969) Measurement of pollutant toxicity to fish. I. Bioassay methods for acute toxicity. *Water Research*, **3**, 793–821.

Sun K, Krause GF, Mayer FL Jr, Ellersieck MR and AP Basu (1995) Predicting chronic lethality of chemicals to fishes from acute toxicity test data: theory of accelerated life testing. *Environmental Toxicology and Chemistry*, **14**, 1745–1752.

Suter GW (1995) Introduction to ecological risk assessment for aquatic toxic effects. In *Fundamentals of Aquatic Toxicology: Effects, Environmental Fate and Risk Assessment*, Rand GM (ed.), Taylor and Francis, Washington, DC, pp. 803–816.

Suter GW, Rosen AE, Linder E and Parkhurst DF (1987) Endpoints for responses of fish to chronic toxic exposures. *Environmental Toxicology and Chemistry*, **6**, 793–809.

Warren-Hicks W, Parkhurst BR and Baker SS (1989) *Ecological Assessment of Hazardous Waste Sites: A Field and Laboratory Reference Document*, EPA 600/3-89/013, Corvalis, OR.

Weibull W (1951) A statistical distribution function of wide applicability. *Journal of Applied Mechanics*, **18**, 293–298.

West Inc. (1994) *Toxstat 3.4*, Western Ecosystems Technology, WY.

Whitehouse P and Cartwright N (1998) Standards for environmental protection. In *Pollution Risk Assessment and Management: A Structured Approach*, Douben PET (ed.), Wiley, Chichester, pp. 235–272.

Whitehouse P, Crane M, Redshaw CJ and Turner C (1996) Aquatic toxicity tests for the control of effluent discharges in the UK – the influence of test precision. *Ecotoxicology*, **5**, 155–168.

Widianarko B and van Straalen N (1996) Toxicokinetics-based survival analysis in bioassays using nonpersistent chemicals. *Environmental Toxicology and Chemistry*, **15**, 402–406.

Zar JH (1984) *Biostatistical Analysis*, 2nd edn, Prentice-Hall, NJ.

10

Trying to Detect Impacts in Marine Habitats: Comparisons With Suitable Reference Areas

A. J. UNDERWOOD

Centre for Research on Ecological Impacts of Coastal Cities, University of Sydney, Australia

10.1 INTRODUCTION

There has been increasing attention to the analysis of pollution and environmental disturbances, partially because substantially better analytical tools and designs are becoming available (e.g. Clarke 1993) and partially because an increasingly sophisticated public is demanding more sensible responses from environmental scientists (e.g. Buhl-Mortensen 1996). There is, however, still a long way to go before the practical problems of environmental sampling will have been resolved satisfactorily.

The major issues for analysis of ecological and environmental sampling depend at least as much (and usually much more) on the framework in which the study is done rather than in the detailed niceties of the statistical procedures to be used. Thus, the scale, scope, timing and type of disturbance or release of contaminants are always going to cause problems for the scientific aspects of a study. It is pointless to collect data, however precise and robust the data are, in the absence of any defined framework for interpreting them. This is well known, but not always acted upon, so some regulatory agencies tend to monitor in the hope that routine measurements without any well-reasoned basis for making them will somehow help solve problems. It is not uncommon for scientific-looking work to be done in response to a sudden calamity (e.g. an oil-spill) without any clearly defined goals.

So, focusing on statistical issues is important, but clarifying the overall aims of the study is more important (Green 1984; Eberhardt and Thomas 1991). The appropriate ways of analysing data are entirely dependent on the hypotheses being tested, the decision-making procedures used to make responses to

Statistics in Ecotoxicology. Edited by T. Sparks. © 2000 John Wiley & Sons Ltd

contaminants or disturbances and on the appropriate choice of variables to measure. Note that these issues remain equally important regardless of whether a logically structured agenda results in tests of clearly formulated hypotheses (e.g. the discussion in Connor and Simberloff 1986; Underwood 1990) or the less clear decision-making procedures favoured by those who find hypothesis-testing distasteful (e.g. Stewart-Oaten 1996; Suter 1996).

The connecting linkage between the two issues – the perceived problem and options for its solution on the one hand and the appropriate statistical analyses on the other – is the choice of sampling and experimental design.

Because problems vary and resources to address them vary even more, the designs of sampling must also vary. Consequently, there is always a need for flexible approaches and tailoring of procedures (see Green 1993). Hence, principles are more important to understand than are particular designs. This makes it difficult to discuss some of the issues by use of a case-history approach (as adopted here), without the details of the cases swamping any more general points. On the other hand, without the details, it may be impossible to understand how the particular study evolved to be as described.

A particular concern and the one considered here is how to incorporate information about reference (or control) locations into an analysis of consequences of contamination or disturbance in one or a series of sites. This is a difficult problem, but various approaches have been used effectively in different studies. Furthermore, there is a large problem in that many environmental issues are a result of accidents or have emerged long after a planned change (see, for example, Wiens and Parker 1995; Glasby 1997). Investigations of these do not have all the logical structure that is provided by a manipulative experiment or by a comparison of what occurs in the disturbed site and suitable reference areas, before and after the disturbance.

In particular, various types of undisturbed reference or control areas have been used in the literature. These have different properties and are, mostly, appropriate for different circumstances. None is ideal because they cannot provide comparisons with natural events or changes in the absence of the environmental disturbance over the period in which it first occurred. Nevertheless, all of the methods imply that there is a common theme about what is required in an analysis of potential environmental consequences of a disturbance. The theme is that the very least data necessary to be able to identify an environmental impact or to measure its magnitude are those enabling comparisons with apparently undisturbed areas.

Documenting differences between potentially impacted and undisturbed reference areas is always difficult because of great variability in time and space (and, usually, different patterns of variability through time in most places) in most of the things that can and should be measured. Nevertheless, interpreting such documented differences is dependent on understanding the limitations imposed by choices of places to sample. The discussion here is to introduce the nature and consequences of such choices.

So, this selective review focuses on cases that reflect the range of methods for selecting references and how to analyse the data after the potential impact has occurred. The examples used were chosen to illustrate different points. There is no attempt to be all-inclusive, nor to attempt a review of the issues currently considered in the literature to be the most pressing. The cases include an analysis of discharges from point-sources into the ocean, each of which was examined by sampling along gradients up and down the coastline. Environmental pollution from an oceanic drilling rig is discussed because it included reference areas in gradients in different directions away from the disturbance. The third case demonstrates methods for using matched pairs of reference and polluted sites in order to extract reliable information about differences, even though there is no proper replication in each pair of sites. The final case considers the use of randomly chosen multiple reference sites to allow comparisons between a polluted and a series of unpolluted places and to allow contrasts between these and the natural differences occurring among unpolluted places.

10.2 SAMPLING DESIGNS FOR COASTAL MARINE POLLUTION

Point-source discharges in coastal waters are one of the abundant types of environmental problems that have received considerable attention. Sampling designs to detect impacts may be based on comprehensive and reliable modelling to indicate the rates of dispersal of contaminants, rates of flow and the capacity of receiving waters to assimilate discharges and the rates of transport (currents, waves, tides) that move contaminants away from the outfall.

If such modelling has been done, sampling can be designed to be very precisely focused on testing hypotheses about rates of dilution and concentrations of contaminants away from the outfall (Figure 10.1). Gradient analysis of some sort is appropriate and can be tailored to the information provided by the modelling. Compliance with criteria set during the design or approval of the discharges can then be assessed with realistic precision.

Alternatively, the issue of concern may be the area over which a particular concentration of contamination is exceeded, rather than the spatial pattern of decline of concentration. For example, a particular chemical (or viruses in sewage) may be discharged which is known to be toxic to humans, if present in the water over a set concentration. It has been predicted from modelling and from the design of the outfall that the specified concentration will only be exceeded within 500 m upstream and downstream of the outfall. Management of this outfall requires regulations to prevent people from swimming in the contaminated area and sampling to test the prediction that concentrations are not being exceeded outside the area.

In this case, sampling does not need to be done to estimate the gradient of concentration away from the outfall (but see below). Instead, sampling effort should be centred at the sites where concentration has been predicted to fall

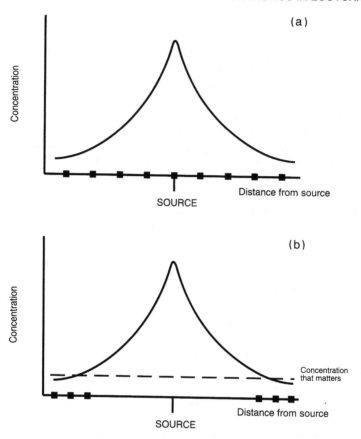

Figure 10.1 Sampling along gradients away from a point-source discharge of contaminants. In (a), modelling, design criteria, etc., have proposed a gradient of decreasing concentration away from the source. Sampling is to test these predictions using sampled estimates of the concentrations at different distances from the source. In (b), modelling or other predictions suggest a gradient of decreasing concentration from the source, but sampling is to test the prediction that concentrations are below the level of concern for management. The solid line indicates the proposed gradient; filled squares indicate appropriate places to sample.

below the set criterion. Further, precautionary management would prevent human access to the water within a further distance – a buffer zone where the amount of contaminant is normally expected to be 'safe'. This buffer zone would provide protection if prior predictions are not exactly correct. Sampling to determine concentrations of contaminants (or viruses) should now be done at the edge of the restricted area (where it is predicted that concentrations are 'safe') and at the edge of the buffer zone (to provide maximal early warnings if things go wrong). The spatial structure of sampling is illustrated in Figure 10.1b.

Note that, if the latter issue (the spatial extent of contamination) is the problem, it is a waste of resources to keep sampling to assess the gradient of concentration. Of course, if the predicted extent of concentration is incorrect and concentrations are excessive further from the outfall than predicted, two responses will be needed. Urgently, management will have to prohibit access to the sea for much greater distances (or close the outfall). Second, sampling to estimate the concentrations of contaminants away from the outfall will be needed. The information can then be used to reassess the licensing of the outfall and to help model the actual dilution and dispersal of discharges.

There are, however, circumstances where sampling at distances away from the outfall is not an appropriate response. For example, it may be known from chemical analysis of the discharges that there is a field of contamination over a set concentration. Outside this, the contaminants are not detectable. The issue of concern may be whether the contamination inside the affected area is having deleterious environmental effects on diversity or structure of assemblages of organisms or some other ecological attribute. In this case, what is needed is a sampling design that allows tests of hypotheses of differences in diversity or assemblages (or whatever relevant variables) between that contaminated area and suitably chosen reference areas. There is no purpose served (there is no relevant hypothesis proposed) that could make use of sampling designed to detect gradients away from the outfall.

10.3 POINT-SOURCE DISCHARGES: SEWAGE ON THE COAST OF NEW SOUTH WALES

Sewage is discharged from the metropolis of Sydney via a number of outfalls (McLean et al. 1991). The three major ones were coastal until 1990 and 1991 when deep-water outfalls were constructed to prevent obvious pollution by untreated sewage on the beaches around Sydney (Fagan et al. 1992). Domestic sewage was discharged with industrial wastes through these three outfalls. Prior to the outfalls being closed and shifted offshore, several studies had been done to determine what contamination they caused and what consequences there may have been for human health. One particular concern in marine systems is the problem of bio-accumulation – the increasing concentration of contaminants in the bodies of marine animals (Phillips 1978).

The example considered here concerns concentrations of organochlorides in tissues of fish (Lincoln Smith and Mann 1989). The fish are known to accumulate contaminants and are consumed by amateur fishermen. The study was motivated by the need to determine whether there was any risk to people eating fish caught at various distances from the outfalls. The study was also designed to compare the three outfalls and to determine concentrations of organochlorides relative to recommended 'safe' levels set by health authorities.

As a result of these requirements, sampling was done at locations spaced at kilometre intervals, starting at 0.5 km each side of each outfall and extending to 3.5 km from the outfall (Figure 10.2).

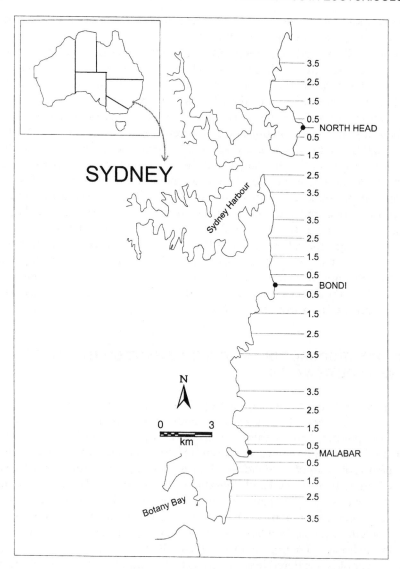

Figure 10.2 Map of coastline of Sydney region, showing sewage outfalls at North Head, Bondi and Malabar. The sites sampled by Lincoln Smith and Mann (1989) at 1 km intervals from 0.5 km north and south of each outfall are shown.

Results (see below) demonstrated that 3.5 km was sufficient for concentrations of contaminants to fall towards background levels. Had this not been the case, the study would have been problematic. If contamination had extended further from the outfalls, the gradient of sampling away from the middle outfall (Bondi) would have become confounded with increasing concentrations from the southern and northern outfalls. Under these circumstances, the entire contaminated area from north of North Head to south of Malabar would have to be considered as a 'field' of pollution. The only relevant gradients that could then have been examined would be northwards from North Head and southwards from Malabar. A quite different study would have been necessary.

Note also that the choice of species to sample prevented taking samples in a seaward direction from the outfalls – the relevant species would not be found in deeper water offshore.

The choice of species to sample is a crucial issue in such a study. There have been several reviews of relevant issues about how to choose an appropriate species or set of species (Underwood and Peterson 1988; Keough and Quinn 1991; Jones and Kaly 1996). Of critical relevance to the present example are the criteria proposed by Hellawell (1986), which include:

1. The species should be easily identifiable, so that taxonomic uncertainty from place to place does not confound differences in concentrations of contaminants between samples.
2. The species should be easy to catch (!).
3. The species should be distributed across the entire area to be sampled, so that the sampling design and planned replication can be achieved.
4. The species should accumulate contaminants; ideally it should be known that rates of accumulation and amounts accumulated truly reflect environmental concentrations.

Finally, for this study of fish, it must be known that the individuals do not move too far during their life, so that their contaminant load is reflective of the area they are caught. Otherwise, all fish in the samples might contain similar concentrations of contamination because they move equally through areas of different concentration.

The red morwong, *Cheilodactylus fuscus*, was chosen by Lincoln Smith and Mann (1989) because it conformed closely to the above criteria. The fish are easy to identify, easy to catch, large enough to provide adequate tissue for assays and available throughout the area. Furthermore, the fish tend to be associated with a particular patch of rocky reef (i.e. they do not move through the whole area). They live for approximately 10 years and feed on a variety of invertebrates (Bell 1979) so are good candidates to accumulate measurable amounts of contaminants, reflective of local concentrations.

Results were very clear-cut. For chlordane (Figure 10.3a), there were obvious excess mean concentrations in fish near each of the outfalls. Centred on each outfall was a peak of concentration, with a decline towards background levels (relative to the level considered acceptable for human consumption according to guidelines by the Australian National Health and Medical Research Council (NHMRC)). South of each of the three outfalls, mean concentrations were still above NHMRC levels at 3.5 km from the outfall. Concentrations were also

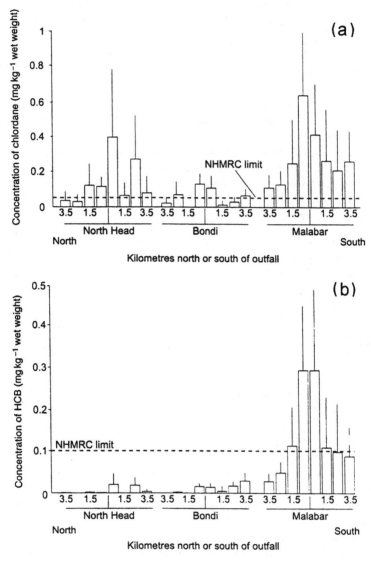

Figure 10.3 Gradients in concentrations of organochlorine contaminants in tissues of a fish at different distances from three sewage outfalls. Data are mean plus 95% Confidence Limits ($n = 8$) and are from Lincoln Smith and Mann (1989) with permission of Dr M. Lincoln Smith.

elevated well to the north of the Malabar outfall. Statistical analyses in Lincoln Smith and Mann (1989) demonstrated that the patterns shown occurred with a small probability of being due to chance. In the second case, elevated concentrations of hexachlorobenzene (HCB) were found only around the Malabar outfall (Figure 10.3b). Concentrations declined as a gradient north and south away from the outfall and mean values were below NHRMC levels about 2.5 km to the north of the outfall.

This example illustrates that gradients in concentrations of contaminants can vary from one outfall to another, in one direction compared with another and from one contaminant to another. Comparative studies of different cases of similar potential impacts are extremely useful (see Underwood 1989 for a discussion of the rationale, and see Olsgard and Gray 1995 for a very fine example comparing numerous oil-rigs).

10.4 POINT-SOURCE POLLUTION AND TWO-DIMENSIONAL GRADIENTS: DISTURBANCES IN SEDIMENTS AROUND AN OIL-RIG

The second case used as an example illustrates a better way of getting control or reference data in an analysis of the environmental effects of a point disturbance. It also demonstrates the utility of an approach using multivariate measures on an entire assemblage of animals. The underlying philosophy of the sampling programme was comprehensively discussed by Field *et al.* (1982) in a pioneering paper on the logic and issues of environmental assessments.

It is worth revisiting the ideal protocols first made explicit in Green's (1979) excellent book and made relevant to multi-species, marine systems by Field *et al.* (1982). Of the various possible procedures for identifying changes or differences in biological assemblages in response to environmental contamination (e.g. Walker *et al.* 1979), Field *et al.* (1982) proposed that it is logical first to sample and analyse for differences in the biota. Then, if there are differences, patterns in the physico-chemical variables may be examined to seek possible explanations for the biological differences. Whether or not this step is necessary is dependent on whether an explanation is actually needed; see Underwood and Peterson (1988) on this point.

The protocol outlined by Field *et al.* (1982) requires a matrix of samples (usually replicated sample units from potentially impacted and reference locations) in each of which the abundances of all taxa are recorded. For this sample × taxon matrix, coefficients are calculated which measure similarity or dissimilarity between every possible pair of sample units. Bray–Curtis measures (Bray and Curtis 1957) are widely used because of their reliability (Faith *et al.* 1987) and recognizable relationships to other measures (Field *et al.* 1982; Clarke 1993).

Calculating the pair-wise coefficients yields a triangular matrix of similarities (or dissimilarities), some of which measure differences between replicate units in each sample (e.g. cores from a potentially impacted site) and some measure differences between units from different sites (i.e. dissimilarity between a core in a reference site and a core in the potentially impacted site). These measures have familiarity as 'within- sample' and 'between-sample' measures in univariate statistical procedures (and see Clarke and Green (1988) for direct parallels). The matrix of dissimilarities can then be classified (through ordination and clustering procedures) to reveal patterns of similarity or difference among samples. These patterns can be analysed by some statistical procedures to indicate probabilities of chance alone being responsible for any differences between contaminated and reference sites.

The example illustrated here is Gray *et al.*'s (1990) analysis of benthic macro-invertebrates subject to possible pollution from the operation of the Ekofisk oil-drilling rig in the Norwegian sector of the North Sea. The detailed methods used were described by Gray *et al.* (1990), but three grab samples of sediment were taken at each of the sampling stations. Individuals of all identifiable taxa (mostly species, some higher taxa) were counted. Counts from each species from the three grabs were summed to give one set of abundances for each site.

The sampling design was very carefully constructed (Figure 10.4). Previous studies indicated that severe environmental impacts occur out to 500 m from a drilling-rig, with some effects detectable out to 1 km (references in Gray *et al.* 1990). Unlike a point-source discharge on a coast with strong along-shore currents or tidal residuals (as in the previous example) and unlike the much

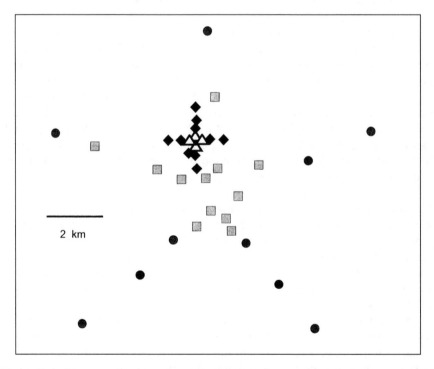

Figure 10.4   Sites sampled around the Ekofisk oil-rig by Gray *et al.* (1990). Stations labelled △ were within 250 m of the rig; ◆, 250–1000 m; ▨, 1000–3500 m; ●, >3500 m; one site not shown was 30 km east of the oil-rig.

simpler linear flows past point-source discharges in rivers (e.g. Faith *et al.* 1995), there is no obvious direction to select to measure gradients of impact away from a drilling-rig. Currents and transport of bottom sediments can be in any direction or vary in direction from time to time.

Therefore, sampling was done on five radii centred on and extending to 4 km from the rig (Figure 10.4). Because of uncertainty over the potential scale of impacts (see later discussion), one sampling station was 30 km away so that it was extremely implausible that this could be affected by the Ekofisk rig. Samples were also concentrated within the innermost 1 km from the rig (16 samples sites were in this inner area).

Ordination of Bray–Curtis similarities was done using ranked data. Thus, the smallest Bray–Curtis measure is assigned the rank of 1, indicating which two sample units are most similar (most like each other). The next smallest Bray–Curtis measure (i.e. next most similar pair) are ranked 2 and so on. Non-metric multidimensional scaling (nMDS) was used to illustrate the ordination. This procedure creates a two-dimensional 'map' of relationships between sample units (three-dimensional maps are also useful). Units are placed on the map so that each is represented as being the right distances from the others (the most similar are closest; the next most similar are the next most close, etc.). Obviously, trying to represent such a complex multidimensional picture in two dimensions creates inaccuracies. The 'stress' of an MDS is a measure of how well the picture represents the original multidimensional pattern. Very small stress (< 0.10) indicates a good to excellent representation; stress < 0.2 is a useable representation (Clarke 1993).

Several groups of similar samples were found, as illustrated by nMDS (Figure 10.5a). Several obvious patterns were revealed. First, there was a striking gradient of structure of assemblages, as indicated by measures of similarity. Samples near the drilling-rig were very different from those further away. Samples about 3 km from the rig, but in any of the directions sampled along the sampled radii were quite similar, clustering together in the nMDS plot (Figure 10.5). This group of samples included the site 30 km away and almost certainly identifies sites that were not impacted by drilling operations.

A second and very important feature of the analysis was that impacts (or, at least, differences from assemblages in uncontaminated reference areas) were evident out to 3 km from the drilling-rig. Some previously used measures (e.g. numbers of individuals per species used by Gray and Pearson 1982) only identified differences in fauna in samples very close to a drilling platform. Measures of diversity and species richness have been recommended for environmental monitoring by Rapport *et al.* (1985) and others. These variables were only different out to 0.5–1 km. The Ekofisk study considerably extended the spatial scale of impacts compared to previous analyses, based on biomass and measures of diversity of animals.

Subsequent to that paper, there have been extensions in the armoury of techniques to unravel specific components of the assemblages and relationships to

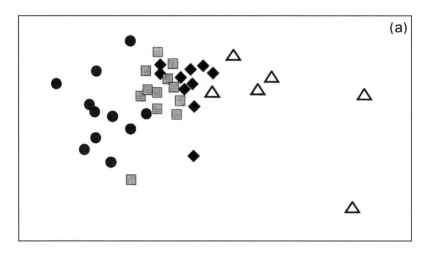

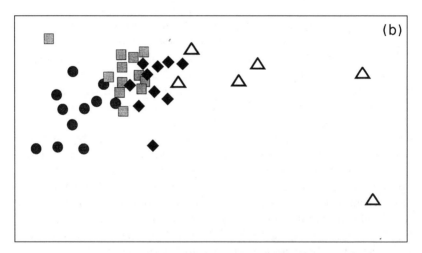

Figure 10.5 nMDS plots illustrating similarities of samples around the Ekofisk oil-rig. Each point represents one site (as in Figure 10.4). Points closer together are more similar. △, < 250 m from the rig; ◆, 250–1000 m; ▨, 1000–3500; ●, > 3500 m. In (a) data from all species show a clear gradient of disturbance at least to the 1000–2500 m region away from the rig (stress = 0.12). In (b), species were grouped into families, making a considerable saving in time, money and taxonomic expertise. Nevertheless, the gradient shown by species (i.e. at greater taxonomic resolution) is still very clear (stress = 0.11). Data supplied by Dr K.R. Clarke.

physico-chemical data. Many procedures were reviewed in detail by Clarke (1993) in a description of permutation procedures for testing for differences among sites (the analysis of similarities or ANOSIM test), to identify which species or taxa contribute most to differences (similarity percentages or SIMPER analysis) and how to do analyses for some particular designs useful in environmental studies. There have been subsequent developments of analysis of correlations between similarity coefficients for the fauna and those for physico-chemical variables (Clarke and Ainsworth 1993; Diniz-Filho and Bini 1996; see Green 1980 for an earlier review).

A particularly stunning finding that has great importance is the discovery that most of the capacity to identify scale of impacts was retained when species were lumped into larger taxonomic units (families, orders, etc.) as illustrated in Figure 10.5b. This has important implications for ecological analyses. It is also very important for the costs of environmental sampling. Much of the expense is in training, infrastructure and access to expertise to identify macrobenthic organisms to species. Where they need to be identified only to larger units, the costs plummet and more resources can be put into taking sufficient samples to ensure adequate power of discrimination between polluted and unpolluted sites.

This is an important point. Taxonomic resolution can be sacrificed without loss of capacity to detect patterns in spatial samples. As a result, a much larger number (or a more complete spatial or temporal coverage) of samples can be taken. Subsequent studies in aquatic habitats have tended to support this finding (Heip *et al.* 1988; Herman and Heip 1988; Warwick 1988a,b; Ferraro and Cole 1990; James *et al.* 1995). Even some studies claiming to demonstrate that fine taxonomic resolution is important demonstrate that most of the spatial differences are preserved with relatively coarse taxonomic groupings (Vanderklift *et al.* 1996).

10.5 AREAS OF CONTAMINATION: FINDING AN EXTERNAL REFERENCE FOR COMPARISONS

Since Green's (1979) pioneering synthesis of the relevant issues, it has become increasingly clear that detecting environmental consequences of contamination cannot rely on a simple 'before-after' contrast of some variable in a particular place, before and then after an environmental disturbance. Thus, even if there are appropriate, relevant and adequately precise data available in some area prior to the arrival of chemical contamination (or other disturbances), demonstrating a difference after the event is not sufficient to be able to claim an environmental impact. Obviously, it is necessary for an impact to be associated with a change from before to after, but this is insufficient evidence because it is potentially confounded by natural changes in the variables measured. It is normal to discover differences in most biological and ecological variables if they are measured at two times separated by several months or years (or longer). Discovering such differences is obviously completely inadequate to allow interpretation that there has been an environmental impact.

In the Bernstein and Zalinski (1983) and Stewart-Oaten *et al.* (1986) paired, before-after methodology (called BACIP), a relevant variable is sampled prior to and after the potential impact in the disturbed area and in a carefully chosen, single control location. Samples are taken at several randomly chosen times prior to the potential impact, but at the same times in the two areas. After the disturbance or contamination starts, sampling is again done at each site at a series of randomly chosen times.

For each time of sampling, there is a sampled difference between the two sites. Prior to the potential impact, the measures of difference are variable from time to time because of natural fluctuations in the measured variable that are not identical in the two sites. If there is no impact, a similar situation prevails after the disturbance and the average difference between the two sites (Stewart-Oaten *et al.* 1986) and the variances of differences (Underwood 1991) should be the same as before.

Using the times of sampling as replicates, the average difference between the two sites can be compared from before to after the disturbance (using a *t*-test or non-parametric equivalent) to test the hypothesis that there was an impact.

There are, of course, numerous problems with this, of which two will be mentioned briefly here. First, there may be some temporal trend in either or both sites in the variable being sampled *before* the disturbance. If so, such trends may not be equal (not present in one site, not in the same direction, not at the same rate) in the two sites. Such prior interaction makes it impossible to predict reliably what amount of change from before to after a disturbance would indicate an impact (Stewart-Oaten *et al.* 1986; Underwood 1992; Smith *et al.* 1993). A realistic solution to this is discussed below (see section 10.6).

The second problem is more profound. The study is, quite simply, un-replicated. There is temporal replication of samples, each of which has some local spatial replication (Stewart-Oaten *et al.* 1986). Despite this, there is no logical basis for concluding that any difference from before to after between one disturbed and one control site represents an impact. This flies in the face of logic (Hurlbert 1984) and most marine and other ecologists' *knowledge*, based on experience with experiments in the field. It is quite usual for biological variables to differ between two sites, to change between two periods of time *and* to change differently in the two sites (see particularly Underwood 1992, 1994a; Smith *et al.* 1993 for discussion).

Therefore, any change that differs between two places is not obviously inter-pretable as an environmental impact – unless it is known to be impossible (or, perhaps, extremely unlikely) naturally. Concentrations of some toxic chemicals are naturally small (or zero). The sudden appearance of unnaturally large amounts of such a chemical in one site where there has been an environmental disturbance would be clear evidence of environmental contamination. Sub-sequent or coincident changes in the fauna, however, could only be interpreted as an effect (an impact) of that contamination if it were known that similar changes do not occur spontaneously in undisturbed sites. This problem can only be solved by analysis of appropriate replicate sites.

Several approaches may be made towards better design. The first considered here extends the methods of station-pairs proposed by Eberhardt (1976) and discussed in a very valuable review by Eberhardt and Thomas (1991). Instead of an analysis that depends on a single comparison between one potentially impacted site and one reference or control site, analyses are made of several such sites.

The example considered here concerns one study of consequences to organisms on rocky shores following the spill of oil from the *Exxon Valdez* (McDonald *et al.* 1993). Numerous sites were available in three regions affected by the spill. In each region, intertidal and subtidal stretches of shore 100–600 km long had been identified that were oiled, or not oiled. An example is given for one region in Figure 10.6. In fact, the particular history of each site was not known exactly because of various attempts to clean oil from some sites. Therefore, any differences between the disturbed and references sites are due to the combination of oil-spill and attempts to clean it up – not just to oil.

Each oiled or unoiled site was classified according to amount of oiling (three

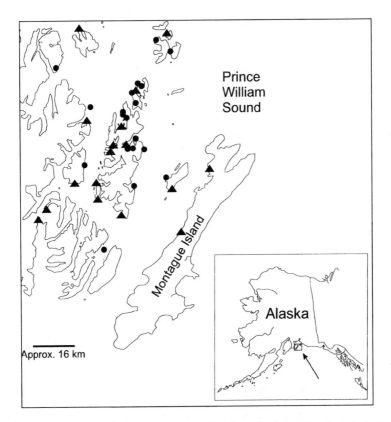

Figure 10.6 Map of Prince William Sound in Alaska showing oiled (●) and unoiled (▲) sites used to analyse effects on barnacles of the oil-spill from the *Exxon-Valdez*. The map was redrawn and modified from one in McDonald *et al.* (1993).

categories, including unoiled) and type of habitat (five types), making 15 categories altogether. The case considered here (i.e. the one described by McDonald *et al.* 1993) is a contrast of moderately oiled and unoiled sites. An unoiled site was chosen which matched as best as possible the physical classification of an oiled site. The variable analysed was the biomass of a particular barnacle, measured in replicated 0.1 m² quadrats in each site (the numbers of such quadrats were not identified in McDonald *et al.* (1993) and their paper made confusing references to *abundances* of barnacles). A two-sample *t*-test was used to compare mean biomasses between the oiled and reference sites (following a Levene's test for homogeneity of variances across sites). It was expected that sites showing effects of oil (and clean-up) would have fewer or smaller barnacles and therefore less biomass. So, a one-tailed *t*-test is appropriate to test the hypotheses:

$$H_0: \mu_r = \mu_o \qquad H_A: \mu_r > \mu_o \qquad (10.1)$$

where μ_r is the mean biomass in the reference site and μ_o is the mean in the oiled site. This null hypothesis would be rejected if the result of the one-tailed *t*-test had small probability, indicating that $\mu_r > \mu_o$, i.e. greater average biomass in the reference site.

Five such comparisons were made (Table 10.1). In all cases, the mean biomasses were greater in the reference sites. If there were no systematic natural differences between the two types of sites, there should be an even chance of the oiled or unoiled site having greater biomass of barnacles. Therefore, the probability of all five pairs having greater biomass in the reference site is (from the binomial distribution) $P = 0.0313$ and can be considered not due to chance, with a small probability of Type 1 error.

For four of the sites, however, the *t*-test was not significant at the conventional (0.05) probability of Type I error. McDonald *et al.* (1993) therefore explored two forms of meta-analysis to combine the outcomes of the five independent tests. First, they used Fisher's (1935) procedure for combining the

Table 10.1. Analysis of biomasses (g per 0.1 m²) of a barnacle on oiled or unoiled shores in Prince William Sound (Alaska). Data from McDonald *et al.* (1993); there may be rounding errors in the calculations done here

Pair of sites	Oiled site		Unoiled site		P-value from t-test	Z-value Z_i	Probability of inclusion I_i	Weighting $W_i = 1/I_i$
	Mean	(SE)	Mean	(SE)				
1	8.36	(6.21)	22.77	(13.18)	0.374	0.321	1.000	1.00
2	2.19	(1.02)	62.81	(32.02)	0.015	2.181	0.262	3.82
3	267.11	(86.24)	428.39	(232.68)	0.167	0.965	0.600	1.67
4	124.78	(23.19)	385.86	(132.77)	0.052	1.630	0.179	5.59
5	66.70	(39.73)	213.71	(112.88)	0.148	1.045	0.667	1.50

probabilities of outcome of each of the tests. If there are k independent tests of an hypothesis (here, $k = 5$) and the probability from the one-tailed t-test in each test is P_i $(i = 1 \ldots k)$, the tests can be combined using:

$$C = -2\sum_{i=1}^{k} \log_e (P_i) \qquad (10.2)$$

where C is distributed as χ^2 with $2k$ d.f. if there are no differences between oiled and unoiled sites (i.e. the null hypothesis is true for all k-tests). The result of this test is $C = 23.68$. The probability of such a large value, by chance, if there were no impacts is $P < 0.01$ (from χ^2 with 10 d.f.).

This analysis also indicates a substantial impact across the five sites. This test does, however, assume that all five sites should have equal importance in determining the outcome and it limits the conclusions to the set of five sites examined. Thus, it is still difficult to generalize to other sites and therefore the entire area. Note also that the test can be significant when a single comparison (one matched pair of sites) gives a very small probability (a highly significant result from that pair's t-test).

The task of the second meta-analysis was therefore to use a test that would reject the null hypothesis (no overall effect of oil-spill/clean-up) when all five component tests are in the same direction and their probabilities are relatively small. The Stouffer–Liptak procedure (Folks 1984) or 'inverse normal' procedure (Rice 1990) has this property.

The test statistic is also useful when different weightings should be placed on the outcomes of the different tests. Here, there is a different chance of each pair of sites being included in the meta-analysis, because different types of habitat (and different amounts of oiling) occurred with different frequencies around the entire area affected by the spill. For each pair of sites selected, it is possible to calculate the probability of inclusion (i.e. the reciprocal of how many sites were available in each category). The inverse of this probability provides a suitable weighting on the grounds that results from a pair with a small probability of being included are, by definition, results from a pair of sites that represent a lot of stretches of shore (i.e. many other possible sites). The probabilities of inclusion of each pair of sites are shown in Table 10.1. Note that the first pair was included because it (uniquely) represented a particular type of site that was considered important. It should therefore have a small 'weight' because it does not represent other pairs. The test statistic is:

$$Z_w = \sum_{i=1}^{k} W_i Z_i \bigg/ \sqrt{\sum_{i=1}^{k} W_i^2} \qquad (10.3)$$

where W_i is the weighting for test i, there are k tests and Z_i is the standard normal score for P_i, the probability associated with the t-test for test i. Thus, Z_i is the standard normal cumulative distribution value corresponding to probability P_i. Z_w itself follows a standard normal distribution and its probability can therefore be found easily. For the data presented by McDonald *et al.* (1993):

$$Z_w = \frac{1.0 \times 0.3 + 3.82 \times 2.18 \ldots + 1.50 \times 1.05}{\sqrt{1.0^2 + 3.82^2 + \ldots + 1.50^2}} = 2.91 \qquad (10.4)$$

with probability, $P < 0.002$.

Again, the conclusion must be that oil and subsequent clean-up are associated with a significant reduction of biomass of barnacles compared with unoiled sites (but see the cautionary comment below).

Clearly, the meta-analyses are all showing a highly significant difference which can (robustly) be attributed to oiling and subsequent clean-up, given that there is so little chance of getting these results if differences between oiled and reference sites were due to some other spatial difference. The results are much more reliable than any possible interpretation from a single oiled and a single reference site.

There are, however, some large problems. Even though the evidence from the various tests indicates very strongly that biomasses of barnacles differ consistently between oiled and unoiled sites, there are two problems for interpreting this to be conclusive evidence that there has been an impact. First, because there are no data for these sites from before the oil-spill, it is not possible to demonstrate conclusively that biomasses have actually changed (decreased or failed to increase relative to what has happened in the reference areas).

Second, because the oil-spill was sudden, large, but unannounced, there is no way to be sure that the very pattern of dispersal of oil is not related to natural differences in biomasses of barnacles. Some patches were oiled. Patches near by were not – presumably because tides, currents, wind differed so that oil missed some, otherwise similar areas. If biomasses of barnacles are themselves influenced by tides, waves, etc. so that biomass is small where certain conditions prevail and these conditions carry oil to such sites, the results indicate only that biomasses differ and so does the amount of oil.

Data in the matched sites from before the spill would have solved the problem. Where data are only available from after the spill, to justify the view that oil caused the differences, it must be demonstrated that the differences found would not occur naturally in the absence of oil. This alternative leads to the hypothesis that, in the absence of an oil-spill, finding matched pairs of sites at the same distances apart, using the same criteria for matching as used in this analysis will not have similar differences of mean biomass as found here. Such matched unoiled sites might be found within the oiled area (Figure 10.6). Alternatively, they could be found in some otherwise similar area of coastline well away from the area directly affected by the oil (although with a spill of the size of that done by the *Exxon Valdez*, that may be impossible).

Nevertheless, the addition of the appropriate 'reference comparison' and discovery that matched pairs of unoiled sites do not differ in biomass of barnacles as much as observed in matched pairs consisting of an oiled and an unoiled site would greatly have strengthened the inference about oil.

10.6 MULTIPLE REFERENCE SITES: BEYOND BACI

An alternative to the meta-analytical approach is to examine more than one impacted and one control site in an orthogonal design (e.g. Underwood 1992; Green 1993). The problem with this is that many environmental problems come about with only one disturbed (and potentially impacted) location. Nevertheless, there is not usually a new or different conceptual problem caused by choosing several reference sites.

This is claimed by some workers to be very difficult – yet they do not find it difficult to choose one reference! The way to establish appropriate references is to consider what you would do if there were several environmentally disturbed sites (e.g. several marinas, outfalls, runways, dredging programmes). These would each be in a site with certain features – sediments, presence or absence of seagrass, near the mouth of an estuary, etc. Whatever these features are, arbitrarily chosen sites with the same features would form the pool of possible reference locations. There is, conceptually, a frequency distribution of sites with varying values of the sorts of variables that constitute the choice of a site to be disturbed (or the sorts of site disturbed by accident).

So, for example, it may be proposed to build a sewage outfall on a rocky headland with a certain minimal depth, currents, wave-action, composition of rock, etc. In a BACIP approach, effort would be spent trying to match these features closely in choosing a reference – in the usually mistaken, but always unsubstantiable belief that biological variables in two locations that *look* similar would continue to behave similarly in the absence of the outfall. It is even possible to measure various attributes in the two sites and to demonstrate that they do not differ. So what? The measurements may not be of the correct variables, may not be powerful enough (Hurlbert 1984) to find important differences (of unspecifiable magnitude) and, anyway, cannot demonstrate that sites will continue to show the same pattern of similarity.

Instead, multiple reference sites should be chosen that provide a sample of sites with general characteristics similar to the one to be disturbed (or that is accidentally disturbed). Note that these sites are not supposed to be identical, but to span the range of variability that exists between the single potentially impacted site and any other (reference) site. These reference sites will provide measures of the magnitudes and nature of temporal changes in the differences between sites in the absence of the environmental problem. From such measures, it is possible to examine the difference between the one potentially impacted site and the reference sites in comparison to the differences among the reference sites. Such comparisons form the basis for tests of hypotheses about impacts (see Smith *et al.* 1993).

A full account of these procedures has been provided elsewhere (Underwood 1991, 1992, 1993, 1994a) and relevant designs for situations where data are only available after a disturbance were discussed by Glasby (1997). There is no room here to reiterate the principles, the practical advantages and the problems. An example of their use, however, comes from the change of disposal of sewage on the coast around Sydney discussed previously.

When the coastal outfalls were decommissioned and sewage was discharged to deeper water offshore, a study was done of the changes in subtidal assemblages around the previous outfalls. The supposition was that organisms in the site immediately around an outfall would be different from those elsewhere because of a long history of exposure to vast volumes of untreated or primary treated sewage. A complex assemblage of encrusting foliose algae and sessile filter-feeders (sponges, ascidians and bryozoans) occupied much of the surfaces on vertical rocky walls around the outfall and in the reference locations (Chapman *et al.* 1995). Four months after the outfall was decommissioned, these assemblages were sampled at three depths (6–9 m, 12–15 m and 18–21 m below mean low water; sewage was discharged 1–2 m below low water before the outfall was turned off). Organisms at different depths were expected to be differently affected by sewage. Two reference locations with similar subtidal cliffs and wave-exposure were sampled 15 km north and 4 km south of North Head.

Six randomly chosen plots of 5 × 3 m separated by about 3 m were sampled at each depth at each location. In each plot, 12 randomly placed photographs (30 × 40 cm) were taken. Of these, six replicates were chosen at random from those where cameras and flashes worked properly and nothing obscured the view. Covers and abundances of identifiable organisms were recorded in each photo.

Several types of analysis were done, using univariate and multivariate methods. Two types of results are illustrated. First, is an example of univariate, asymmetrical ('beyond BACI', see section 10.6) analysis of the number of types of sponges (Table 10.2). The rationale for this type of analysis is explained in full in Underwood (1992, 1993, 1994a). In principle, the differences among the three locations can be separated into two components. The first is the difference between the putatively polluted location and the average at the two reference areas. Second is the difference between the two reference areas. The first of these measures the actual impact, in the sense that it determines the probability of error in concluding that a polluted site is different from the average at the unpolluted reference areas. The second component examines the natural variability between undisturbed reference areas. Environmental impacts can only be detected, or their magnitudes measured against natural variation.

In the analysis in Table 10.2, there are also interactions between depth and the differences among locations. Of these, the component 'Depth × Outfall versus References' identifies any difference, from one depth to another, in the pattern of difference between mean number of sponges at the outfall and in the

Table 10.2 Asymmetrical ('beyond BACI') analysis of the mean number of types (different taxa) of sponges in six randomly chosen plots at three depths near the North Head outfall ('O' or Outfall locations) and two reference locations (R_1 and R_2); $n = 6$ replicate quadrats in each plot (from Chapman *et al.* 1995).

Source of variation	d.f.	Mean Square	F-ratio
Among Locations	2		
$\quad$ O versus References	1	37.08	
$\quad$ R_1 versus R_2	1	0.78	
Among Depths	2		
Depth $\times$ Locations	4		
$\quad$ Depth $\times$ O versus References	2	45.48	2.24 $P > 0.05$
$\quad$ Depth $\times$ R_1 versus R_2	2	20.28	3.66 $P < 0.01$
Plots (Locations)	45		
$\quad$ Plots (Depth $\times$ O)	15	1.57	0.94 $P > 0.05$
$\quad$ Plots (Depth $\times$ References)	30	5.53	3.31 $P < 0.001$
Residual	270	1.67	

references areas. Thus, if any influences of sewage cause a difference in the number of species of sponges, but effects are not similar at different depths, this interaction would be large. The other component (Depth $\times$ R_1 versus R_2) measures the natural (i.e. in the absence of an outfall) amount of interaction – the extent to which natural differences between the two reference areas vary with depth.

Finally, there are components which measure the spatial variability among plots in each depth and location. Again, there are two components of these. First is variability among plots in the different depths of the outfall location (there are six plots and therefore 5 d.f. for comparing their means, at each of three depths; hence 15 d.f. in Table 10.2). Second is variability among plots at the three depths in the two reference locations (5 d.f. for the six plots at three depths for two locations; a total of 30 d.f. in Table 10.2).

The outcome of the analysis (and subsequent multiple comparisons) for the number of types of sponges was quite clear. There was significant variation among plots at the various depths at the outfall location (i.e. Plots(Depth $\times$ References) in Table 10.2). This implies that there is more variability among plots (i.e. at the spatial scale of several metres) at the outfall location than is normally found in undisturbed areas.

The average number of sponges per replicate photo also differed between two reference areas, but in different ways at different depths (Depth $\times$ R_1 versus R_2), but there was no significant difference between the outfall and average of the reference locations (Table 10.2). The details are discussed in Chapman *et al.* (1995) and Underwood and Chapman (1997).

The sampling design also allowed tests of hypotheses that spatial variability will be altered by pollution around the outfall. This is important because some authors consider increased variability among samples may be a general response to pollution (Warwick and Clarke 1993). Also, in many marine habitats, patchiness in occupancy of space is considered important for maintaining diversity (e.g. Levin and Paine 1974; Paine 1974; Connell and Keough 1985).

In the example discussed here, the variances among quadrats in each plot provided six replicate estimates of variation (at a scale of a few metres in each plot) for each depth and at each location. These were analysed as ratios of variance to mean for each plot, to remove influences of trends of increasing variance with mean. The ratios were transformed to logarithms to normalize the data. The two-factor analysis of variance had locations and depths as orthogonal factors, with six replicate plots in each combination (details in Chapman *et al.* 1995).

Variability from one plot to another can also be examined to test hypotheses about influences of sewage on patchiness at the scale of approximately 10 m between centres of plots. Some examples of the results of these analyses are shown in Table 10.3. In the first two cases (Table 10.3a and b), there were substantial differences between the outfall and the reference areas. Mean cover of encrusting algae was much less variable among plots at the outfall location than at the reference locations; mean density of an anemone was considerably

Table 10.3 Examples of analyses of variability among replicate plots ($n = 6$ plots) at various depths in the location with an outfall (O) and two reference locations (R_1 and R_2). Differences were examined by two-tailed F-ratios (with 5 and 5 d.f.); nominal probabilities of Type I error for each set of tests were adjusted to $P = 0.05$ by the Bonferroni procedure to help account for there being three tests in each set. ('>' or '<' indicates a difference at $P = 0.05$; '=' indicates no significant difference); data supplied by Dr MG Chapman

		Variance among plots		
		O	R_1	R_2
(a)	Percentage cover of encrusting algae: 18–21 m			
		0.42	252.25	914.67
	Outcome:	$O < R_1$; $O < R_2$; $R_1 = R_2$		
(b)	Density of anemone, *Anthothoe albicincta*: 6–9 m			
		21.48	0.76	0.55
	Outcome:	$O > R_1$; $O > R_2$; $R_1 = R_2$		
(c)	Density of tubeworms: 18–21 m			
		0.31	0.015	0.044
	Outcome:	$O > R_1$; $O > R_2$; $R_1 < R_2$		
d)	Percentage cover of hydroids: 6–9 m			
		0.70	0.03	15.48
	Outcome:	$O > R_1$; $O < R_2$; $R_1 < R_2$		

more variable than at the reference areas. Both imply a previous effect of sewage on the patchiness of the organisms.

The latter two cases illustrate the importance of having more than one reference area. Variation in mean density of tubeworms differed between the two reference locations, but both had significantly smaller variance than at the outfall (Table 10.3c). Patchiness at the outfall was outside the range of values for areas unaffected by sewage.

In contrast, variation among patches in the percentage cover of hydroids (Table 10.3d) in the shallowest depth differed among all three locations. The reference locations differed and the location with the outfall fell within this range of natural variation. This provides no evidence for an impact due to the outfall.

Note that had there been only one reference location (as so strongly recommended by devotees of the BACIP procedure), the differences between either O and R_1 or O and R_2 (depending upon which was chosen) would have been interpreted as an impact. Unless the amount of difference that occurs naturally from place to place has already been estimated, it is impossible to interpret sensibly any difference between two places, one of which happens to have an outfall.

10.7 DISCUSSION

The cases discussed indicate the range of possibilities for comparing potentially impacted to reference areas. In some circumstances, it is appropriate to consider a gradient (in one or two dimensions) away from the source of contamination or disturbance. There are two interesting features of such analyses relevant to this discussion.

First, it is extremely important to be sure that an appropriate scale of sampling has been chosen to ensure that enough ground is covered so that impacted areas (if they exist) are really being compared to unimpacted areas. Many studies attempt to determine the relevant scales of sampling appropriate for a particular impact by modelling or monitoring physical and chemical variables (e.g. Spellerberg 1991).

This leads to several possible issues. There may be no good correlation between physical and chemical variables on the one hand and biological or ecological variation on the other. Thus, contamination (the presence of unnatural chemicals) is not the same as pollution (the effects of or responses to such contamination). Also, many marine organisms filter large quantities of water during respiration or feeding and are therefore exposed to considerably larger quantities of chemicals than occur in relatively small samples of water for assays. Organisms also accumulate toxins, so that, if their toxicity is only manifest over a certain threshold concentration, it will be realized in organisms living in water that is apparently not toxic.

Further, the scale of ecological variation in assemblages (e.g. abundances, varieties, structure) is often poorly matched to the scale of sampling of chemistry in seawater. It is not surprising, therefore, to find mismatches between gradients of chemistry and gradients of ecology. It is, however, more peculiar, but quite realistic to find mismatches between gradients of chemistry and physics. As an example, Raimondi and Reed (1996) examined numerous physical, chemical and ecological variables away from the discharge from diffusers of process water on the Californian coast. Only three of seven physical or chemical variables were correlated with distance from the diffusers. In contrast, magnitudes of 12 out of 16 biological variables were correlated with distance.

Different assemblages and different species in an assemblage will often respond differently to a chemical insult, so biological effects are not necessarily easily predicted as simple gradients from a discharge. Measures of water quality are not necessarily useful surrogates for or indicators of ecological impacts.

So, predicting scales of impacts is much less sure than being able to measure them. Clearly, in many studies, data are collected before a disturbance and the design of sampling is based on modelling and other predictions. Such studies will be severely compromised if the scale of modelling is wrong so that the scales of sampling were incorrectly anticipated.

Keough and Black (1996) pointed out that many assessments of environmental impacts are over small spatial scales (<10 km), although the oil-spill from the *Exxon Valdez*, as one example, was very much greater than this. In many marine habitats, being comfortable that the scale of sampling is adequate requires that predictions about the scales of potential impacts include explicit information about scales of influence of larval dispersal into and out of the potentially impacted sites. If an impact decimates the breeding population of some species in a relatively small area, there may be effects over very large distances because densities of planktonic dispersive larvae will be reduced. Subsequent recruitment into many widespread patches of habitat may therefore be reduced (although actual magnitudes of such reduction will be small and the power to detect them very little). The converse is also possible; a patch of habitat contaminated by chemicals may be quite large, but recruitment of new organisms may continue at normal rates because they are arriving from unaffected populations elsewhere (Underwood and Peterson 1988).

Getting the scale of sampling right is an extremely important, although difficult, requirement. Designs that can cope with several nested (or hierarchical) scales will be better than anything which assumes that the scale is known (see particularly Underwood (1992) on this point and for an example of hierarchical analysis).

If the scale of sampling is completely mismatched by being much smaller than the scale of effects, there are very serious consequences. For example, the ecological assemblages out to several (say > 3) kilometres from an outfall may all be altered in similar ways by contamination in the discharges. If prior expectations have led to sampling out to 2 km, with the last half-kilometre considered unaffected (i.e. a suitable reference), no impact can possibly be detected. All samples will appear to be similar to those in the reference area.

This may seem an obvious point, but has been misused in some studies (as discussed in Underwood 1994b). It is, in fact, highly desirable for polluters (who presumably wish not to be caught) to pay only for sampling over too small (or the wrong) scale.

The second major issue about using gradients concerns the way they are sampled. Studies involving gradients tend not to be concerned with careful 'matches' or attempts to match sites. In most cases, what drives the choice of sampling sites is distance from the source. This is in stark contrast to the care to attempt to match a single reference site with a single disturbed site in BACIP procedures (Stewart-Oaten et al. 1986). As stated earlier, the process of matching sites assumes that all relevant variables have been identified before stratification is done to match the site. It is further assumed that these variables are of appropriate magnitude to provide a match (which implies that the link between these variables and the biological or ecological variables to be sampled are understood). Finally, it is assumed that, as time passes, the supposedly matched reference site would continue to match the disturbed area, if it were not for the presence of an impact to the disturbed site.

Sampling along gradients and use of multiple reference areas (Underwood 1992, 1994a) share the first assumption. In analyses of gradients, however, there is less difficulty than in a paired study about the exactness of a 'match'. Several sites are sampled at different distances from the disturbance, so variation among sites is inevitably assessed in the analyses of the data. The second of the above assumptions is not an issue for studies with multiple reference areas, because there is no pretence that there is a match with the disturbed site. Instead, there is the assumption that a sample has been taken of possible matching sites from a population of such matching sites.

The problem of there continuing to be a match between a reference and a disturbed location is a less serious issue when sampling is done over several times after the disturbance. Such temporal replication or analysis of temporal trends is extremely important (Green 1979; Stewart-Oaten et al. 1986; Underwood 1991). There is no such problem where multiple controls or references are used (Smith et al. 1993; Underwood 1994a) because the degree to which sites change relative to each other under natural conditions and temporal interactions among sites are explicitly measured and can be contrasted to differences between reference and disturbed sites (see Glasby 1997).

Obviously, any comparisons of a disturbed site to reference sites to detect impacts and to estimate their magnitude should be based on as much information as possible. Inclusion of multiple reference sites and, if appropriate, analyses of gradients across multiple references will help a lot in most circumstances.

As a final point, without data from before a disturbance, it is impossible to be very sure about an impact (Green 1979; Stewart-Oaten et al. 1986). The whole notion of a disturbance causing an impact is intimately tied to the demonstration that the effect occurs in the disturbed site (not in reference areas) and following the disturbance (not before it). This really requires data from before to after and from the disturbed site (if there must be only one) and several replicated references. Then conclusions about impacts can be fairly based on long-standing principles of evidence for causation (Mill 1865).

Temporal sampling brings with it a whole set of problems about relevant time-scales for detecting short- or long-term responses (pulse or press impacts) to short- or long-term (pulse or press) disturbances. There is no place here to consider these issues, but consult Bender *et al.* (1984), Underwood (1991), Glasby and Underwood (1996) for some discussion.

It will always be better to have replicated, independently interspersed assignment of disturbed and reference areas throughout a study site and adequate data from before and after the disturbance (to control the experiment properly). In the absence of a truly controlled experiment (which would still have problems about deciding the appropriate temporal and spatial scales), sampling in one disturbed and several reference locations before and after a disturbance is still the preferable option. Often, however, a comparative study must be made that can only use information from after the disturbance. Where and how to get the information must, as illustrated here, be thought about extremely carefully before sampling starts. Clear aims, clear goals, thought about spatial (and temporal) scales and methods of analysis will all increase the suitability of the sampling and therefore the worth of any subsequent interpretations.

ACKNOWLEDGEMENTS

The preparation of this paper was supported by funds from the Australian Research Council and the Institute of Marine Ecology (of the University of Sydney). I thank Dr MG Chapman for advice, discussion and for some unpublished data and Ms V. Mathews for help with the preparation of the figures.

REFERENCES

Bell JD (1979) Observations on the diet of the red morwong, *Cheilodactylus fuscus* Castelnau (Pisces: Cheilodactylidae). *Australian Journal of Marine and Freshwater Research*, **30**, 129–133.

Bender EA, Case TJ and Gilpin ME (1984) Perturbation experiments in community ecology: theory and practice. *Ecology*, **65**, 1–13.

Bernstein BB and Zalinski J (1983) An optimum sampling design and power tests for environmental biologists. *Journal of Environmental Management*, **16**, 335–343.

Bray JR and Curtis JT (1957) An ordination of the upland forest communities of Southern Wisconsin. *Ecological Monographs*, **27**, 325–349.

Buhl-Mortensen L (1996) Type-II statistical errors in environmental science and the precautionary principle. *Marine Pollution Bulletin*, **32**, 528–531.

Chapman MG, Underwood AJ and Skilleter GA (1995) Variability at different spatial scales between a subtidal assemblage exposed to the discharge of sewage and two control assemblages. *Journal of Experimental Marine Biology and Ecology*, **189**, 103–122.

Clarke KR (1993) Non-parametric multivariate analyses of changes in community structure. *Australian Journal of Ecology*, **18**, 117–143.

Clarke KR and Ainsworth M (1993) A method of linking multivariate community structure to environmental variables. *Marine Ecology Progress Series*, **92**, 205–219.

Clarke KR and Green RH (1988) Statistical design and analysis for a 'biological effects' study. *Marine Ecology Progress Series*, **46**, 213–226.

Connell JH and Keough MJ (1985) Disturbance and patch dynamics of subtidal marine animals on hard substrata. In *Natural Disturbance: the Patch Dynamics Perspective*, Pickett STA and White PS (eds), Academic Press, New York, pp. 125–151.

Connor EF and Simberloff D (1986) Competition, scientific method and null models in ecology. *American Scientist*, **75**, 155–162.

Diniz-Filho JAF and Bini LM (1996) Assessing the relationship between multivariate community structure and environmental variables. *Marine Ecology Progress Series*, **143**, 303–306.

Eberhardt LL (1976) Quantitative ecology and impact assessment. *Journal of Environmental Management*, **42**, 1–31.

Eberhardt LL and Thomas JM (1991) Designing environmental field studies. *Ecological Monographs*, **61**, 53–74.

Fagan P, Miskiewicz AG and Tate PM (1992) An approach to monitoring sewage outfalls: a case study on the Sydney deepwater sewage outfalls. *Marine Pollution Bulletin*, **25**, 172–180.

Faith DP, Dostine PL and Humphrey CL (1995) Detection of mining impacts on aquatic macroinvertebrate communities: results of a disturbance experiment and the design of a multivariate BACIP monitoring programme at Coronation Hill, Northern Territory. *Australian Journal of Ecology*, **20**, 167–180.

Faith DP, Minchin PR and Belbin L (1987) Compositional dissimilarity as a robust measure of ecological distance. *Vegetatio*, **69**, 57–68.

Ferraro SP and Cole FA (1990) Taxonomic level and sample size sufficient for assessing pollution impacts on the Southern California Bight macrobenthos. *Marine Ecology Progress Series*, **67**, 251–262.

Field JG, Clarke KR and Warwick RM (1982) A practical strategy for analysing multi-species distribution patterns. *Marine Ecology Progress Series*, **8**, 37–52.

Fisher RA (1935) *The Design of Experiments*, Oliver and Boyd, Edinburgh, 244 pp.

Folks JL (1984) Combination of independent tests. In *Handbook of Statistics 4. Nonparametric Methods*, Krishnaiah PR and Sen PK (eds) North-Holland, New York, pp. 72–94.

Glasby TM (1997) Analysing data from post-impact studies using asymmetrical analyses of variance: a case study of epibiota on marinas. *Australian Journal of Ecology*, (**22**, 448–459).

Glasby TM and Underwood AJ (1996) Sampling to differentiate between pulse and press perturbations. *Environmental Monitoring and Assessment*, **42**, 241–252.

Gray JS, Clarke KR, Warwick RM and Hobbs G (1990) Detection of the initial effects of pollution on marine benthos: an example from the Ekofisk and Eldfisk oilfields, North Sea. *Marine Ecology Progress Series*, **66**, 285–299.

Gray JS and Pearson TH (1982) Objective selection of sensitive species indicative of pollution-induced change in benthic communities. I. Comparative methodology. *Marine Ecology Progress Series*, **9**, 111–119.

Green RH (1979) *Sampling Design and Statistical Methods for Environmental Biologists*, Wiley, Chichester, 257 pp.

Green RH (1980) Multivariate approaches in ecology: the assessment of ecological similarity. *Annual Review of Ecology and Systematics*, **11**, 1–14.

Green RH (1984) Statistical and non-statistical considerations for environmental monitoring studies. *Environmental Monitoring and Assessment*, **4**, 293–301.

Green RH (1993) Application of repeated measures designs in environmental impact and monitoring studies. *Australian Journal of Ecology*, **18**, 81–98.

Heip C, Warwick RM, Carr MR, Herman PMJ, Huys R, Smol N and Van Holsbeke K (1988) Analysis of community attributes of the benthic meiofauna of Frierfjord/Langesundfjord. *Marine Ecology Progress Series*, **46**, 171–180.

Hellawell JM (1986) *Biological Indicators of Freshwater Pollution and Environmental Management*, Elsevier Applied Science, London, 546 pp.

Herman PMJ and Heip C (1988) On the use of meiofauna in ecological monitoring: who needs taxonomy? *Marine Pollution Bulletin*, **19**, 665–668.

Hurlbert SJ (1984) Pseudoreplication and the design of ecological field experiments. *Ecological Monographs*, **54**, 187–211.

James RJ, Lincoln Smith MP and Fairweather PG (1995) Sieve mesh size and taxonomic resolution needed to describe natural spatial variation of marine macrofauna. *Marine Ecology Progress Series*, **118**, 187–198.

Jones GP and Kaly UL (1996) Criteria for selecting marine organisms in biomonitoring studies. In *Detecting Ecological Impacts: Concepts and Applications in Coastal Habitats*, Schmitt RJ and Osenberg CW (eds), Academic Press, San Diego, pp. 29–48.

Keough MJ and Black KP (1996) Predicting the scale of marine impacts: understanding planktonic links between populations. In *Detecting Ecological Impacts: Concepts and Applications in Coastal Habitats*, Schmitt RJ and Osenberg CW (eds), Academic Press, San Diego, pp. 199–234.

Keough MJ and Quinn GP (1991) Causality and the choice of measurements for detecting human impacts in marine environments. *Australian Journal of Marine and Freshwater Research*, **42**, 59–554.

Levin SA and Paine RT (1974) Disturbance, patch formation, and community structure. *Proceedings of the National Academy of Sciences*, **71**, 2744–2747.

Lincoln Smith MP and Mann A (1989) *Bioacumulation in Nearshore Marine Organisms II: Organochlorine Compounds in the Red Morwong, Cheilodactyus fuscus, Around Sydney's Three Major Sewage Ocean Outfalls*, The Ecology Lab, Sydney, 25 pp.

McDonald LL, Erickson WP and Strickland MD (1993) Survey design, statistical analysis, and basis for statistical inference in coastal habitat injury assessment: *Exxon Valdez* oil spill. In *Exxon Valdez Oil Spill: Fate and Effects in Alaskan Waters*, Wells PG, Butler JN and Hughes JS (eds), American Society for Testing and Materials, Philadelphia, pp. 296–311.

McLean C, Miskiewicz AG and Roberts EA (1991) Effect of three primary treatment sewage outfalls on metal concentrations in the fish *Cheilodactylus fuscus* collected along the coast of Sydney, Australia. *Marine Pollution Bulletin*, **22**, 134–140.

Mill JS (1865) *A System of Logic, Volume 2*, 6th edn, Longmans, Green, London, 553 pp.

Olsgard F and Gray JS (1995) A comprehensive analysis of the effects of offshore oil and gas exploration and production on the benthic communities of the Norwegian continental shelf. *Marine Ecology Progress Series*, **122**, 277–306.

Paine RT (1974) Intertidal community structure: experimental studies on the relationship between a dominant competitor and its principal predator. *Oecologia*, **15**, 93–120.

Phillips DJH (1978) The use of biological indicator organisms to quantitate organo-chlorine pollutants in aquatic environments: a review. *Environmental Pollution*, **16**, 167–229.

Raimondi PT and Reed, DC (1996) Determining the spatial extent of ecological impacts caused by local anthropogenic disturbances in coastal marine habitats. In *Detecting Ecological Impacts: Concepts and Applications in Coastal Habitats*, Schmitt RJ and Osenberg CW (eds), Academic Press, San Diego, pp. 179–198.

Rapport DJ, Regier HA and Hutchinson TC (1985) Ecosystem behavior under stress. *American Naturalist*, **125**, 618–640.

Rice WR (1990) A consensus combined *P*-value test and the family-wide significance of component tests. *Biometrics*, **46**, 303–308.

Smith EP, Orvos DR and Cairns J (1993) Impact assessment using the before-after-control-impact (BACI) model: concerns and comments. *Canadian Journal of Fishery and Aquatic Science*, **50**, 627–637.

Spellerberg IF (1991) *Monitoring Ecological Change*. Cambridge University Press, Cambridge, 334 pp.

Stewart-Oaten A, Murdoch WM and Parker KR (1986) Environmental impact assessment: 'pseudoreplication' in time? *Ecology*, **67**, 929–940.

Stewart-Oaten A (1996) Goals in environmental monitoring. In *Detecting Ecological Impacts: Concepts and Applications in Coastal Habitats*, Schmitt RJ and Osenberg CW (eds), Academic Press, San Diego, pp. 17–28.

Suter GW (1996) Abuse of hypothesis testing statistics in ecological risk assessment. *Human and Ecological Risk Assessment*, **2**, 331–347.

Underwood AJ (1989) The analysis of stress in natural populations. *Biological Journal of the Linnean Society*, **37**, 51–78.

Underwood AJ (1990) Experiments in ecology and management: their logics, functions and interpretations. *Australian Journal of Ecology*, **15**, 365–389.

Underwood AJ (1991) Beyond BACI: experimental designs for detecting human environmental impacts on temporal variations in natural populations. *Australian Journal of Marine and Freshwater Research*, **42**, 569–587.

Underwood AJ (1992) Beyond BACI: the detection of environmental impact on populations in the real, but variable, world. *Journal of Experimental Marine Biology and Ecology*, **161**, 145–178.

Underwood AJ (1993) The mechanics of spatially replicated sampling programmes to detect environmental impacts in a variable world. *Australian Journal of Ecology*, **18**, 99–116.

Underwood AJ (1994a) On beyond BACI: sampling designs that might reliably detect environmental disturbances. *Ecological Applications*, **4**, 3–15.

Underwood AJ (1994b) Things environmental scientists (and statisticians) need to know to receive (and give) better statistical advice. In *Statistics in Ecology and Environmental Monitoring*, Fletcher D and Manly BJ (eds), University of Otago Press, Dunedin, pp. 33–61.

Underwood AJ and Chapman MG (1997) Subtidal assemblages on rocky reefs at a cliff-face sewage outfall (North Head, Sydney, Australia): what happened when the outfall was turned off? *Marine Pollution Bulletin*, **33**, 293–302.

Underwood AJ and Peterson CH (1988) Towards an ecological framework for investigating pollution. *Marine Ecology Progress Series*, **46**, 227–234.

Vanderklift MA, Ward TJ and Jacobi CA (1996) Effect of reducing taxonomic resolution on ordinations to detect pollution-induced gradients in macrobenthic infaunal assemblages. *Marine Ecology Progress Series*, **136**, 137–145.

Walker HA, Saila SB, and Anderson EL (1979) Exploring data structure of New York Bight benthic data using post-collection stratification of samples, and linear discriminant analysis of species composition comparisons. *Estuarine and Coastal Marine Science*, **9**, 101–120.

Warwick RM (1988a) Analysis of community attributes of the macrobenthos of Frierfjord/ Langesundfjord at taxonomic levels higher than species. *Marine Ecology Progress Series*, **46**, 167–170.

Warwick RM (1988b) The level of taxonomic discrimination required to detect pollution effects on marine benthic communities. *Marine Pollution Bulletin*, **19**, 259–268.

Warwick RM and Clarke KR (1993) Increased variability as a symptom of stress in marine communities. *Journal of Experimental Marine Biology and Ecology*, **172**, 215–226.

Wiens JA and Parker KR (1995) Analyzing the effects of accidental environmental impacts: approaches and assumptions. *Ecological Applications*, **5**, 1069–1083.

Appendices

Appendix 1 Great skua data. Reprinted from *Environmental Pollution*, Vol. 19, R Furness and M Hutton. Pollutant levels in the great skua, pages 261–268, Copyright 1979, with permission from Elsevier Science (* Not determined)

Tissue data			Mercury		Cadmium		Selenium	
	bird	Primary feather	Kidney	Liver	Kidney	Liver	Kidney	Liver
1	3	*	10.4	15.0	64.6	8.8	27.9	23.0
2	4	4.1	*	3.2	*	6.8	*	9.1
3	4	10.3	7.9	6.3	13.5	1.0	19.3	18.3
4	4	3.8	3.4	3.6	50.0	4.3	*	*
5	*	3.9	4.9	4.9	52.6	4.3	25.5	18.4
6	5	11.1	10.1	*	14.5	*	30.0	*
7	7	1.8	4.2	6.5	129.2	8.2	13.3	6.7
8	7	11.5	18.6	27.6	21.6	2.0	33.0	30.6
9	8	3.4	6.1	3.6	29.7	1.8	20.8	7.1
10	8	2.9	4.1	4.5	58.6	6.5	*	*
11	9	2.0	6.3	9.6	114.7	10.8	*	23.3
12	10	4.0	19.5	9.3	91.2	4.4	36.6	26.0
13	12	15.1	*	30.4	336.0	31.4	89.1	34.6

Egg data

Row	PCB	DDE	DDD	DDT	Dieldrin	HCB	Mercury	Cadmium
1	19	1.7	0.04	0.12	0.160	0.04	0.275	0.022
2	13	1.1	0.05	0.16	0.092	0.04	0.284	0.002
3	10	1.6	0.05	0.22	0.073	0.03	0.409	0.015
4	33	3.2	0.07	0.22	0.088	0.09	1.311	0.002
5	10	1.2	0.04	0.10	0.079	0.03	0.575	0.012
6	36	3.6	0.07	0.28	0.150	0.16	0.349	0.002
7	19	2.5	0.10	0.55	0.089	0.07	0.086	0.013
8	30	1.7	0.05	0.24	0.086	0.06	0.609	0.002
9	13	1.4	0.04	0.15	0.056	0.05	0.713	0.011
10	6	0.4	0.03	0.05	0.022	0.03	0.302	0.018
11	33	3.2	0.04	0.27	0.096	0.07	0.304	0.016
12	21	1.7	0.11	0.56	0.097	0.08	0.472	0.018

Appendix 2   Beetle behaviour data, courtesy of Peter Hankard, ITE Monks Wood

Animal number	Copper dose	Turning rate	Turning bias	Path length (mm)	Velocity (mm s^{-1})	Number of stops	Active time (s)
1	0	3.45	0.67	2206	8.53	118	259
2	0	6.93	0.54	6022	12.19	913	494
3	0	5.99	−0.09	2076	11.80	401	176
4	0	6.85	0.53	2626	11.61	504	223
5	0	9.09	0.20	766	5.03	241	118
6	0	5.57	−0.44	4124	11.87	426	347
7	0	7.45	−0.14	5913	10.04	898	582
8	0	8.36	−0.46	5865	9.84	1008	596
9	0	6.26	−0.18	5593	10.67	767	524
10	0	8.57	−0.01	4855	10.11	828	480
11	0	5.58	0.44	19498	10.64	2107	1832
12	0	8.09	−0.62	2326	9.93	419	234
13	0	6.03	0.05	2163	11.94	419	181
14	0	6.21	0.18	2563	11.83	489	217
15	0	4.75	0.08	8321	13.14	2045	630
16	0	3.74	0.03	2876	12.99	624	221
17	0.5	6.78	−0.13	97	11.56	25	8
18	0.5	2.12	−0.17	185	11.46	45	16
19	0.5	4.66	0.68	662	12.58	187	53
20	0.5	5.54	−0.26	4409	12.34	1134	357
21	0.5	4.17	−0.04	20015	13.56	4239	1476
22	0.5	4.42	0.12	8870	7.91	2414	785
23	0.5	5.54	−0.26	4409	12.34	1134	357
24	0.5	4.17	−0.04	20015	13.56	4239	1476
25	0.5	4.42	0.12	8870	7.91	2414	785
26	0.5	4.22	−0.02	1992	13.29	535	150
27	0.5	4.58	−0.25	1816	13.25	448	137
28	0.5	4.05	−0.85	595	12.41	173	48
29	0.5	6.25	0.19	18292	11.18	2649	1636
30	0.5	6.06	0.26	4636	10.34	719	448
31	0.5	4.10	−0.17	2176	12.90	589	169
32	0.5	4.74	−0.10	715	12.29	213	58
33	0.5	3.77	−0.20	2491	13.37	627	186
34	0.5	7.42	−0.07	2078	9.94	339	209
35	1	0.00	0.00	23	8.16	15	3
36	1	6.74	0.82	541	11.85	177	46
37	1	4.63	0.46	601	12.94	149	46
38	1	3.82	−0.12	9053	13.17	1858	687
39	1	3.94	−0.26	5486	12.91	1282	425
40	1	4.32	0.20	3971	13.14	982	302
41	1	3.91	−0.31	9817	9.19	415	1068
42	1	6.57	0.51	360	4.56	47	79
43	1	3.80	0.24	5477	13.40	1359	405
44	1	4.07	−0.17	11604	13.13	2936	882
45	1	4.02	−0.10	16897	12.83	3927	1281
46	1	4.77	0.14	2425	12.49	673	189
47	1	4.44	−0.42	3090	13.09	765	236

Appendix 2   *Continued*

Animal number	Copper dose	Turning rate	Turning bias	Path length (mm)	Velocity (mm s^{-1})	Number of stops	Active time (s)
48	1	4.97	0.69	1897	12.60	521	151
49	1	5.49	0.08	1113	12.87	303	86
50	1	0.00	0.00	3	3.90	2	1
51	1	4.13	−1.12	899	12.41	276	72
52	1	8.25	−0.44	4736	9.81	903	476
53	1	7.59	0.18	13087	10.61	1828	1232
54	1	7.47	0.62	2382	10.24	366	232
55	1	6.03	0.01	2958	10.00	557	296
56	1	9.18	−0.33	1798	9.72	307	185
57	2	3.81	0.73	925	13.39	234	69
58	2	3.84	−0.30	2588	13.47	509	192
59	2	3.87	0.28	1659	13.14	406	126
60	2	4.59	0.22	24945	12.28	3393	2004
61	2	5.34	0.02	15772	10.03	2798	1396
62	2	4.88	0.46	2380	12.48	619	190
63	2	2.77	−0.14	1418	14.06	312	101
64	2	0.00	0.00	9	6.68	4	1
65	2	4.72	−0.03	3158	12.47	862	253
66	2	5.11	0.10	876	12.79	224	68
67	2	5.16	−0.34	3556	12.61	948	282
68	2	3.73	0.97	249	13.07	54	19
69	2	4.91	2.02	366	13.39	89	27
70	2	4.81	0.03	1990	12.58	535	156
71	2	6.02	0.32	1480	12.62	373	117
72	2	3.90	0.11	4370	13.99	864	312
73	2	3.90	0.31	4570	12.91	1084	348
74	2	5.30	−0.35	6032	12.58	1636	479
75	2	4.79	1.05	508	13.93	100	36
76	2	5.42	−0.02	1595	12.54	423	127
77	2	5.15	−0.03	6529	12.47	1072	523
78	2	5.00	0.05	11272	12.25	1882	920
79	2	4.68	−0.03	14376	12.17	2062	1170

Appendix 3 Starling data, courtesy of Dan Osborn, ITE Monks Wood. All metal determinations are in mg kg⁻¹ dry weight

Sequential bird no.	Month 1=Jan, 13=Jan	1 male, 2 female	Moult score	Liver wt (g)	Lipid in liver (mg g⁻¹)	Liver lipid load (g)	Liver protein conc (mg g⁻¹)	Protein load (g)	Body lean dry wt (g)	Body lipid (g)	Pectoral muscle lean dry weight (g)	Zn liver	Cu liver	Fe liver	Ca liver	Mn liver	Cd liver	Hg liver
1	1	2	36	4.27	98.15	0.419	173	0.739	27.32	19.58	4.90	68.55	15.65	2376	169.1	4.48	1.35	1.49
2	1	1	36	4.60	54.34	0.250	129	0.593	24.22	14.29	3.91	68.91	18.65	1460	154.1	2.95	1.95	2.39
4	1	2	36	5.00	120.00	0.600	143	0.715	24.13	16.51	3.65	52.24	12.53	1401	177.9	2.71	2.08	0.08
5	2	1	36	4.65	49.59	0.231	87	0.405	25.33	15.08	4.39	72.92	18.72	1841	166.5	5.93	4.40	0.90
6	2	2	36	3.45	10.10	0.035	103	0.355	21.37	13.14	3.91	75.02	15.81	2018	141.7	4.65	2.45	0.08
7	2	1	36	4.10	55.56	0.228	110	0.451	26.66	13.34	4.95	68.64	15.81	2332	224.9	5.49	2.78	0.08
8	2	1	36	3.45	24.39	0.084	105	0.362	24.18	19.32	4.59	64.64	17.47	3501	175.1	3.74	2.27	0.48
9	4	2	36	3.89	99.17	0.386	112	0.435	23.77	3.85	4.24	67.80	16.75	565	127.0	3.51	2.30	0.08
10	4	2	36	4.24	67.50	0.286	109	0.462	22.94	4.41	4.19	64.51	16.41	564	219.2	4.52	2.53	0.66
11	4	2	36	3.64	63.29	0.230	105	0.382	23.62	4.59	4.00	89.57	20.81	1712	589.4	4.73	3.39	0.66
12	4	1	36	2.75	89.55	0.246	170	0.467	23.80	3.55	4.37	84.60	22.17	4014	105.6	4.65	3.30	0.27
13	4	2	36	3.63	71.23	0.256	125	0.454	25.24	3.91	4.11	57.78	23.60	3574	91.59	2.72	1.95	0.28
14	4	2	36	3.66	12.61	0.046	144	0.527	24.11	3.33	4.40	49.41	13.19	860	187.0	3.41	2.46	0.08
15	5	2	36	2.68	23.53	0.063	127	0.340	21.67	0.76	3.88	103.30	24.19	2989	305.9	4.83	2.91	0.18
16	5	1	36	2.53	31.91	0.081	126	0.319	24.42	2.23	4.09	87.24	21.47	1315	161.2	5.43	5.16	0.08
17	5	2	36	2.25	23.25	0.052	150	0.333	22.37	0.93	3.30	107.20	33.03	1272	269.4	3.10	4.93	0.08
18	5	2	36	2.63	9.35	0.025	98	0.257	22.94	2.01	3.66	100.20	24.70	1096	188.8	3.52	11.38	0.31
19	6	2	36	4.11	46.15	0.189	72	0.296	20.68	2.89	3.43	60.77	15.19	1076	111.6	2.53	2.93	0.08
20	6	2	36	4.65	55.56	0.258	78	0.363	21.75	3.56	3.74	63.72	13.36	1166	109.4	3.13	2.56	0.08
21	6	1	1	4.05	50.00	0.202	107	0.433	23.40	2.86	3.79	73.59	13.72	1376	107.9	5.78	5.12	0.08
22	6	1	36	4.28	59.26	0.254	104	0.455	23.68	1.66	3.93	61.29	18.33	1263	100.7	3.08	4.81	0.08
23	7	1	27	2.89	41.67	0.120	108	0.312	21.51	3.91	3.46	76.24	20.35	4817	440.2	6.44	4.17	0.08
24	7	1	24	3.23	13.15	0.042	100	0.323	23.38	3.31	3.90	64.50	13.50	3369	184.5	4.44	3.45	0.08
25	7	1	20	2.55	28.57	0.073	99	0.252	23.47	2.77	3.57	88.89	14.20	1755	130.5	5.26	7.39	0.08
26	7	1	20	2.55	19.61	0.050	94	0.400	21.53	2.12	3.35	93.90	17.91	3111	203.7	5.28	5.58	0.08
27	7	1	27	4.44	22.10	0.098	78	0.346	23.82	3.42	3.83	71.18	14.72	3146	127.7	5.22	2.41	0.26
28	7	2	24	3.00	56.34	0.169	92	0.276	24.33	3.72	3.73	84.95	16.93	2792	559.6	5.65	7.29	0.78
29	9	1	36	3.50	42.25	0.148	123	0.431	27.10	3.05	4.21	76.80	12.28	4243	329.6	5.98	5.31	0.08
30	9	1	35	3.75	56.07	0.210	99	0.371	27.87	2.67	4.71	106.00	21.85	4389	121.4	0.47	9.23	0.08

Appendix 3 *Continued*

Sequential bird no.	Month 1=Jan, 13=Jan	1 male, 2 female	Moult score	Liver wt (g)	Lipid in liver ($mg\,g^{-1}$)	Liver lipid load (g)	Liver protein conc ($mg\,g^{-1}$)	Protein load (g)	Body lean dry wt (g)	Body lipid (g)	Pectoral muscle lean dry weight (g)	Zn liver	Cu liver	Fe liver	Ca liver	Mn liver	Cd liver	Hg liver
31	9	2	36	3.75	48.19	0.180	122	0.457	22.07	4.12	3.64	76.50	16.85	3798	245.0	3.20	4.62	0.08
32	9	1	36	3.59	27.78	0.094	148	0.500	24.79	2.74	3.89	65.58	19.64	2356	130.8	5.17	4.82	0.08
33	9	1	35	4.09	70.00	0.286	97	0.396	27.50	3.11	4.69	101.50	20.30	6156	174.2	4.21	6.62	0.79
34	8	1	26	5.30	98.36	0.521	105	0.557	24.29	3.86	3.97	61.48	10.52	1756	158.1	3.16	3.49	0.08
35	8	2	20	3.36	28.57	0.096	96	0.322	22.05	2.65	3.82	78.25	23.48	1569	176.2	2.63	6.06	0.08
36	8	2	16	3.01	101.50	0.305	103	0.310	21.95	3.42	3.50	66.14	15.88	2633	172.1	4.95	5.92	0.08
37	8	1	27	3.74	6.85	0.026	78	0.291	23.25	2.77	3.70	72.78	19.12	3826	221.4	4.53	3.76	0.08
38	8	2	26	3.70	47.50	0.176	99	0.366	22.54	3.50	3.84	58.43	9.06	3066	142.5	4.47	4.28	0.08
39	10	1		3.78	100.80	0.381	94	0.355	26.31	10.81	4.29	55.80	15.21	3477	172.5	3.66	2.82	0.08
40	10	1		3.79	90.91	0.344	89	0.377	24.73	2.53	4.32	97.60	18.40	2902	437.5	12.15	4.55	0.08
41	10	2		3.42	100.00	0.342	100	0.342	27.37	3.99	4.65	100.40	16.05	5115	401.4	9.03	6.10	0.08
42	10	1		3.70	104.80	0.388	106	0.392	28.09	2.90	4.54	57.22	18.60	2783	179.2	3.83	4.09	0.08
43	10	2		4.40	21.28	0.094	93	0.409	29.31	4.94	4.73	80.32	24.65	3757	379.5	5.44	8.12	0.08
44	11	1		4.21	105.80	0.445	96	0.404	23.99	4.25	3.69	58.88	16.18	2501	125.1	3.25	2.02	2.38
45	11	1		4.10	67.96	0.278	96	0.393	24.13	5.78	3.64	67.89	17.67	2883	149.4	3.50	4.37	1.55
46	11	2		4.09	141.70	0.580	88	0.360	22.38	5.38	3.63	74.70	22.76	2310	227.6	6.85	3.59	12.1
47	11	2		4.05	29.41	0.119	95	0.384	23.86	4.46	3.85	67.26	16.08	3044	152.2	3.90	2.73	0.23
48	11	1		4.17	107.70	0.449	106	0.441	25.76	5.01	3.88	74.00	18.16	2303	160.4	4.62	3.05	1.09
49	12	1		3.89	101.35	0.394	94	0.365	25.73	8.35	3.77	97.13	19.20	4537	304.9	8.70	3.56	7.05
50	12	1		4.02	56.06	0.225	123	0.494	27.13	9.36	4.22	100.40	16.70	3447	543.9	14.32	3.06	8.62
51	12	1		4.53	82.35	0.373	125	0.566	26.17	13.22	4.07	84.41	18.61	2481	535.8	8.62	3.45	1.69
52	12	1		4.04	140.58	0.567	130	0.525	25.86	13.40	4.02	89.79	17.96	3049	407.0	10.17	2.13	0.74
53	12	1		4.24	109.09	0.463	100	0.424	26.44	6.65	3.88	84.43	16.10	3103	268.3	6.81	1.89	1.96
54	13	1		3.51	92.44	0.324	163	0.572	23.22	13.77	3.97	86.58	21.01	2328	576.1	6.83	2.56	0.67
55	13	2		4.01	120.96	0.484	141	0.565	24.52	6.95	3.92	65.95	13.67	1527	112.7	2.75	0.79	0.08
56	13	2		4.80	106.12	0.509	95	0.465	23.62	10.63	4.23	75.96	19.30	3619	225.9	4.24	1.47	0.55
57	13	2		4.47	113.00	0.505	123	0.550	26.37	11.24	4.18	69.61	12.17	1848	178.3	5.11	2.93	0.08
58	13	2		3.85	42.55	0.164	111	0.427	23.40	8.57	3.90	63.92	23.34	2775	229.9	6.07	3.73	6.39

Appendix 4 Streamwater chemistry data from 25 Welsh plantation sites of different ages. Data courtesy of Paul Stevens, ITE Bangor

Age	NO$_3$N	pH	Na	K	Ca	Mg	Al	Cl	SO$_4$S	Si	DOC	DON	Mn	Fe
0	0.13	4.91	4.36	0.14	0.68	0.57	0.105	8.18	1.17	0.38	1.27	0.06	0.03	0.01
0	0.15	5.00	3.91	0.21	0.68	0.74	0.148	7.44	1.43	0.46	1.54	0.02	0.02	0.03
0	0.05	5.37	3.98	0.16	1.15	0.79	0.091	7.30	1.22	0.75	5.61	0.15	0.04	0.23
0	0.11	5.27	4.24	0.16	0.98	0.74	0.096	7.93	1.23	0.70	2.61	0.05	0.01	0.03
0	0.10	6.95	3.84	0.22	3.39	1.91	0.031	7.78	1.45	0.90	2.26	0.05	0.01	0.02
10	0.05	4.44	4.32	0.15	0.63	0.69	0.193	7.89	1.22	0.78	8.66	0.12	0.04	0.26
14	0.10	4.67	4.65	0.09	0.53	0.71	0.301	8.61	1.57	0.81	1.53	0.03	0.02	0.03
14	0.02	5.34	4.89	0.08	1.44	0.81	0.053	9.07	1.44	0.80	2.82	0.08	0.19	0.22
15	0.14	5.96	4.36	0.21	1.39	1.07	0.018	8.82	1.34	0.77	0.69	0.01	0.01	0.01
16	0.07	4.31	5.10	0.93	0.74	0.83	0.196	9.61	1.46	0.79	17.14	0.11	0.06	1.13
19	0.07	4.58	4.91	0.11	1.25	0.73	0.149	9.32	1.70	0.78	5.41	0.11	0.31	0.24
24	0.12	5.09	6.19	0.09	1.70	0.83	0.232	11.57	1.69	1.08	3.41	0.07	0.03	0.05
27	0.12	5.57	8.05	0.31	1.94	1.57	0.042	15.49	2.34	1.18	0.57	0.00	0.01	0.01
28	0.33	5.19	4.15	0.12	0.76	0.60	0.097	7.44	1.06	0.87	0.56	0.01	0.01	0.01
29	0.14	4.83	6.10	0.19	1.40	0.77	0.174	11.14	1.57	0.90	2.99	0.04	0.04	0.05
31	0.20	5.37	5.32	0.17	1.82	1.08	0.074	9.58	1.93	1.63	4.08	0.10	0.01	0.07
32	0.16	4.93	5.39	0.14	1.17	0.94	0.216	10.17	1.95	1.05	4.40	0.05	0.16	0.30
37	0.64	4.74	6.67	0.19	1.07	1.21	0.432	12.17	2.01	1.53	1.08	0.01	0.02	0.04
40	0.33	5.13	5.42	0.24	1.38	0.95	0.166	10.16	1.98	0.94	0.47	0.01	0.03	0.01
44	1.11	4.61	10.42	0.20	0.73	1.36	1.020	19.51	2.03	0.75	0.76	0.04	0.11	0.02
51	0.26	4.40	7.95	0.21	0.73	1.03	0.619	14.23	2.14	0.88	3.22	0.04	0.03	0.09
53	0.36	4.75	5.99	0.22	1.08	1.03	0.412	11.05	2.26	0.97	0.88	0.03	0.06	0.03
53	0.39	5.24	5.92	0.23	1.61	1.13	0.193	10.87	2.19	1.14	0.70	0.02	0.03	0.02
53	0.59	5.38	7.20	0.24	2.85	1.91	0.093	14.16	2.88	1.17	0.55	0.02	0.02	0.01
55	0.94	4.63	8.77	0.23	1.80	1.19	0.597	16.51	2.43	0.60	1.21	0.04	0.13	0.02

Index